Quantum Courtroom: The Trial of a Century

Sandeep Chavan

Published by SANDEEP CHAVAN, 2025.

QUANTUM COURTROOM: THE TRIAL OF A CENTURY

First edition. July 3, 2025.

ISBN: 979-8231381913

Written by Sandeep Chavan.

Table of Contents

To the misread thinkers, the silenced equations, and the brave few who dare to realign science with reality.

"Truth was never hidden.It simply waited—until perception aligned."— *Universal Energy Dynamics*

Disclaimer

This book presents an alternative scientific and philosophical perspective based on the author's independent framework—Universal Energy Dynamics (UED). It is not intended to replace any legally accepted scientific curriculum, institutional theory, or verified experimental standard.

The content challenges conventional interpretations of quantum mechanics, perception, and physical laws. While the ideas are rooted in rational inquiry and consequence-based logic, they reflect a paradigm shift that may diverge sharply from mainstream views.

Readers are encouraged to engage critically and openly, with a spirit of curiosity—not dogma.

This work is not affiliated with any academic body, government agency, or professional scientific institution. It is a personal intellectual contribution intended to provoke thought, invite debate, and expand the scope of inquiry in science and philosophy.

All references to historical figures, concepts, and theories are made with deep respect, even when critiqued.

This is not the final truth. It is a ripple. What you do with it determines the next alignment.

Preface

Why We Brought Quantum to Trial

For over a century, we've accepted a version of reality written in the cryptic language of mathematics, wrapped in paradox, and narrated by equations that could not answer their own questions.

We were told that the universe was fundamentally uncertain, that particles could be in multiple places at once, and that nothing was real until we looked at it. And when we found contradiction, we made peace with it—not by resolving it, but by romanticizing it.

This book is not a rebellion. It is a **reckoning.**

Quantum Courtroom was born out of one deep realization: *we were never looking at reality itself—we were witnessing our misalignment with it.* What we called collapse, randomness, duality, or entanglement were not properties of nature but consequences of our unreadiness, our lack of synchrony with the field we inhabit.

With the Universal Energy Dynamics (UED) framework, we don't discard quantum theory—we realign it. We bring back the observer, not as a passive measurer, but as a fundamental participant in perception itself. We expose frequency as perception timing, constants as thresholds, and particles as illusions of consequence.

This is not a textbook. It is a **trial**—one where equations are interrogated, thought experiments are cross-examined, and legends of science are called back to the witness stand. It is bold, theatrical, and unapologetically disruptive.

But it is also humble—because it does not claim to replace the great minds. It simply listens more closely to the ripples they heard, but misread.

If this book succeeds in one thing, let it be this: To remind us that reality was never hiding. It was **waiting**—for us to align.

– Sandeep Chavan

Author, Ripple Thinker, Consequence Aligner

How to Read This Book

Not as a textbook. Not as a theory. But as a trial.

This book is structured like a courtroom drama, not a physics manual. Every chapter is a witness, every concept is on trial, and you—the reader—are part of the jury.

Here's how to approach it:

Start with Curiosity, Not Conclusions

Don't worry if you're not a physicist. This isn't about academic qualification—it's about your willingness to question what you've been told is unquestionable.

Each Chapter Is a Standalone Statement

You can read this book sequentially or jump to any section. Every chapter carries a complete argument, yet together they form a powerful verdict.

Keep the Glossary Handy

Terms like ATT, ripple alignment, and field memory are new lenses. Refer to the Glossary (Appendix A) when needed. You'll soon begin to see with ripple vision.

Forget the Wave–Particle Debate

You're not here to settle old science. You're here to step beyond it—into a framework where consequence, perception, and intent rule the field.

Don't Look for Math—Look for Meaning

This book speaks the language of consequence, not calculus. It doesn't model—it realigns.

Pause Often. Reflect Deeply. Align Internally.

Some chapters may feel disruptive. That's not discomfort—it's recalibration. Let it happen.

The Last Parts Are Forward-Looking

Parts IV and V aren't critiques—they are blueprints for the future. If you're an educator, technologist, thinker, or student, you'll find your ripple there.

This Is Not the End. It's the Opening Argument.

Read once. Then again. Let the field speak to you through silence, not just words.

Because in the end, the book doesn't close the case. It hands you the gavel.

Prologue: Before the Gavel Fell

It was never meant to come to trial. The equations were beautiful. The interpretations—elegant. And for a time, physics walked with the pride of a prophet whose predictions matched the stars.

But behind the precision of probabilities and the glittering abstractions of particles and waves, something quietly fractured. The closer we got to the truth, the further it drifted into paradox. The more we measured, the less we understood. And at the heart of it all—an innocent cat, a collapsing wave, and a universe that seemed to only make sense *when we looked at it*.

This is not a courtroom in Vienna or Geneva. This is a tribunal of consequences—where thought itself stands trial. Where Schrödinger, Heisenberg, Bohr, Born, and Dirac are not condemned, but called— to face the alignment they missed.

This book is not a rejection of quantum theory. It is a cross-examination of how we misunderstood it.

We didn't fail because we lacked intelligence. We failed because we assumed that the universe must conform to our methods—that equations dictate existence, that time flows like a river, that probability governs reality, and that observation collapses truth into form.

But now, something clearer is rippling through. A model where perception is not passive, where the field is not inert, and where reality doesn't "decide" what to be—but responds when alignment is ready.

This is the courtroom where we question not what the universe is—but why we insisted it was something else.

Welcome to Quantum Courtroom. The trial has begun.

Part I – Opening the Case: The Math That Misled

1: Opening Statement

The Case Against Collapse

The Charges: Mathematical Absolutism, Observer Neglect, and the Duality Myth

Ladies and gentlemen of the rational world, We are here today not to destroy quantum physics, but to liberate it from misinterpretation.

We stand not against the work of giants, but against the blind allegiance to their shadows. We do not indict Schrödinger, Heisenberg, or Dirac. We indict the collapse—of reason, of coherence, of truth—disguised as a wave function reduction.

What began as a brilliant exploration into the unknown has become a shrine of assumptions. And today, we bring it to court.

Exhibit 1: Mathematical Absolutism

They told us the universe is written in the language of mathematics. They said if it cannot be quantified, it cannot be real. They built an empire of equations, polished with symbols, and wrapped it in certainty.

But what they forgot—math doesn't create reality. It describes it.

A Reminder of What Math Is:

- Math is a system of symbols.
- It is a language of abstraction, not existence.
- It does not pulse, ripple, breathe, or observe.
- It follows us—we do not follow it.

When we turned Ψ into a deity and Schrödinger's equation into scripture, we began worshiping the model—not the mechanics behind it.

We replaced ripple dynamics with probability distributions.

We ignored that every mathematical description is built upon perceptual consequences—and that these consequences arise only when alignment between field and observer occurs.

In other words:

Math is a record of resonance. Not the music itself.

Exhibit 2: Neglect of the Observer

Perhaps the gravest of sins in quantum interpretation is treating the observer as a nuisance—an afterthought.

We wrapped the observer in uncertainty, blurred them with probabilities, and buried them beneath measurement devices. But the observer was never outside the system. They were the key to it.

The observer is not an intruder. The observer is the very mechanism by which reality gets written.

Quantum theory tried to resolve this through Copenhagen, through Many Worlds, through decoherence—but each was a patch. None of them asked the real question: "Why does the field respond only when the observer is ready?"

That readiness is not philosophical fluff. It's quantifiable through ATT—Alignment Threshold Time.

If there is no alignment, there is no consequence. If there is no consequence, there is no perception. And if there is no perception—reality remains unexpressed.

This isn't mysticism. It's the most honest science we've refused to do.

Exhibit 3: The Myth of Duality

The wave-particle duality was a placeholder—an admission of confusion that somehow became a dogma.

We didn't understand what we were observing, so we created a paradox:

- Light is a wave.
- Light is a particle.
- It depends on whether you look.

What kind of science makes what something is depend on how you look at it?

Let us be precise:

- A wave is not a thing.
- A particle is not a thing.
- Both are perceptual outcomes of ripple alignment.

What we call a particle is simply a localized consequence of ripple resolution. What we call a wave is the periodic perception of unresolved alignment.

There is no duality. There is only ripple choreography—the interplay of field tension and observer intent.

The Collapse: A Fictional Event

The wave function never collapsed. It was never waving to begin with.

What collapsed was our alignment with the field—the moment it responded to the observer's readiness. What we called "collapse" was the resolution of ripple potential into a consequence.

Schrödinger never claimed Ψ was real. He knew it was a predictive map. It was Born who gave it probabilities, and Bohr who gave it mystery.

But no one gave it a mechanism—until now.

A Universe That Never Needed Dice

Einstein famously said, "God does not play dice." Bohr told him to stop telling God what to do.

But they were both wrong—and right.

God doesn't play dice. And the universe doesn't need to.

There is no randomness—only uncalibrated alignment. There is no unpredictability—only unknown readiness.

When field and observer meet at ATT, consequence is born. And that moment was always bound to happen. It just waited for you to be aligned enough to see it.

Your Honor, the Case Is Clear

We charge quantum orthodoxy with the following:

1. **Mathematical Absolutism**: Declaring equations as

reality rather than consequence.

2. **Observer Neglect**: Treating perception as interference instead of participation.
3. **The Duality Myth**: Calling one outcome two identities because we refused to understand alignment.

We are not here to destroy science. We are here to restore it—to its consequential roots, to its relational truths, and to a reality where ripple dynamics replace paradoxes.

Closing Insight:

"The wave function didn't collapse. It resolved because you were finally aligned. The cat was never in danger. Only the meaning was."

2: Exhibit A

The Fallacy of Frequency

"Frequency was never emitted. It was how often we aligned."

In the theater of physics, few words carry as much mathematical gravity as frequency. It is carved into Planck's law, drives the heartbeat of electromagnetic theory, and measures the rhythm of everything from a violin string to a quantum oscillator. It seems fundamental.

But today, we place it on the stand.

Because what if frequency—like time, like waves, like particles—is not something the universe emits...but something we construct from the rhythm of our own perception?

Let the record show:

Frequency was never emitted. It was how often we aligned.

What Science Told Us

In classical physics, frequency is clean:

- $f = 1/T$ — the number of cycles per second.
- In waves: the number of oscillations per unit time.
- In quantum mechanics: frequency links directly to energy via $E = hf$.

This frequency is said to belong to the source:

- The atom emits a photon at a certain frequency.
- A star pulses in measurable cycles.
- A microwave hums with predictable wavebands.

We built entire fields on this foundation. And yet—we never asked the deeper question:

"How do we know the frequency is real—and not just when we noticed it?"

The Observer's Hidden Clock

Enter ATT—Alignment Threshold Time.

In the UED framework, ATT defines the minimum time interval required for a ripple-field interaction to register as perception. It's not a mechanical tick. It's a readiness threshold—a physiological, psychological, and energetic gateway.

Let's reframe:

- A source doesn't *emit* a frequency.
- It continuously exists as a ripple strain in the field.
- You perceive it in rhythmic pulses based on how often your own alignment threshold is crossed.

Thus:

- You don't receive a wave.
- You receive ripple confirmations at fixed intervals.
- Those intervals feel periodic—and we call that frequency.

The so-called "frequency" is just your interaction rate with the ripple field.

Frequency as Perception, Not Emission

This is a massive shift. Let's clarify the consequences.

Classical View:

- A source vibrates at 500 THz.
- A wave is emitted at that frequency.
- You detect it as light.

UED View:

- A source introduces ripple tension into the structured field.
- If you are aligned, you resolve this ripple periodically.
- The rate at which you resolve it = perceived frequency.

Thus, two observers with different field alignment conditions might detect:

- Different frequencies.
- Different coherence.
- Even no detection at all, though the ripple tension still exists.

That's not relativity. That's alignment dependency.

The Doppler Deception

Let's put the Doppler Effect on the stand.

Traditionally:

- A moving source compresses or stretches the wave.

- This changes the frequency observed (redshift/ blueshift).

But what's really happening?

The *source* isn't changing its emission. Your alignment conditions are changing.

- The ripple field doesn't carry a wave.
- It carries structured strain, which is resolved differently depending on relative motion, field tension, and observer readiness.

So Doppler isn't about frequency being warped in transit. It's about alignment intervals shifting due to motion-induced phase dissonance.

The Great Equation Trap

Physicists say: "But we *measure* the frequency. It's real."

Yes, you measure how often something registers—in your calibrated instruments. Those instruments have their own perceptual ATT thresholds (clock speeds, sampling rates, sensor limits).

This is still perception—not universal emission.

Imagine:

- A rotating lighthouse flashes every 5 seconds.
- Two observers, differently located and angled, both measure "5 seconds"—but the mechanism never flashed anything.
- It continuously radiated ripple tension.
- What they "saw" was when their field alignment met

that ripple.

That 5-second interval? Not the source's rhythm. The observer's reception window.

Rewriting E = hf

This sacred equation, too, must be revisited.

In UED:

- h remains a calibration constant.
- But f is not emitted—it's resolved.
- So E = hf is not "the energy of the photon" but the energy resolution rate of the observer at a given ATT alignment frequency.

Thus:

Energy is not in the packet. It's in the consequence of ripple alignment.

Closing Arguments: Ripple, Not Rhythm

Let us return to the charge:

The Fallacy of Frequency is the belief that:

- Frequency is an objective emission.
- The wave carries rhythm.
- The observer merely detects it.

The truth, now revealed:

- Frequency is not a signature of the source.
- It is a rhythm of observer-field interaction.
- It is not sent—it is resolved.

Closing Insight:

"The universe didn't pulse. You did. What you called a frequency was just your rhythm of meeting the ripple."

3: Witness Called – Schrödinger

"His equation wasn't a wave. It was a field of unresolved possibilities."

The bailiff calls out: *"The Court summons Erwin Schrödinger, theoretical physicist, author of the famous wave equation and reluctant father of a dead-and-alive cat."*

Schrödinger steps forward—not as a ghost of reputation, but as a ripple of consequence, formed by the unfinished understanding of his own legacy.

Opening Question:

Sandeep (Cross-Examiner): "Professor Schrödinger, let's go straight to the point. When you formulated your equation, did you imagine a real wave physically propagating in space?"

Schrödinger (Calmly): "No. I described a mathematical construct, a tool to calculate the likelihood of a system's measurable outcomes over time. Not a vibrating entity. Ψ was not real—it was predictive."

Misread from the Start

Schrödinger's equation was elegant:

$$i\hbar\frac{\partial}{\partial t}\Psi(x,t) = \hat{H}\Psi(x,t)$$

It looked like a wave equation. It felt like a wave equation. So they treated Ψ like a wave, with peaks and troughs and interference patterns—something physically sloshing through space.

But in truth:

- Ψ was never meant to describe a particle's motion through space.
- It was meant to describe a system's state of potential—a field of what could be, not what is.

Schrödinger himself grew uncomfortable with the literal interpretation of Ψ. He never embraced the idea of a particle being "smeared out" across space like a ghost. His cat was a critique—not a model. A parody of the absurdity.

"I don't like it, and I'm sorry I ever had anything to do with it." — Schrödinger (historical letter)

What Ψ Actually Represents (UED Reframe)

Under UED, the Schrödinger equation doesn't model a physical wave. It models a field of unresolved ripple alignments—a probabilistic cloud *only because* alignment has not yet occurred.

In clearer terms:

- Ψ is the map of ripple possibilities across the structured vacuum field.

- The field isn't waving—it's awaiting.
- The moment RAC (Ripple Alignment Criterion) + ATT (Alignment Threshold Time) are met—one of those ripple possibilities is resolved.

This moment, classically called "wave function collapse," is redefined in UED as:

Ripple Resolution Event (RRE) — where potential field alignment becomes consequential observation.

Superposition: The Intellectual Mirage

The infamous superposition principle—where a particle exists in all states until observed—was never Schrödinger's belief. It was a misapplied extrapolation from linear algebra into ontology.

In UED, there is no particle "existing in all states." There is only a field of unexpressed ripple alignments.

Once the observer's field condition satisfies the minimum ATT, a single outcome is rendered. Not chosen, not collapsed—resolved.

The cat was not both dead and alive. It was never a cat in a box at all. It was a metaphor—meant to shock us into questioning what we were really modeling.

Collapse: The Fantasy That Followed

Bohr and Born took the Ψ function and made it mystical:

- Born gave it probabilistic meaning: $|\Psi|^2$ as likelihood.
- Bohr created a fog around measurement: complementarity, observer effect, duality.

What no one provided: a mechanism.

They called it "collapse." But what collapsed? Nothing. There was no mechanism—only *mystery posing as maturity*.

Collapse became the band-aid for a question no one wanted to answer: What actually happens between field potential and perception?

The Ripple Clarification

UED gives us that missing mechanism.

- The field is not filled with probabilities.
- It is filled with ripple patterns, each seeking resolution.
- The observer's presence doesn't collapse anything—it aligns.
- When alignment occurs (RAC + ATT), a perception event happens.
- What appears as "particle behavior" is simply a localized consequence of this ripple resolution.

Closing Dialogue (Ripple Courtroom Style)

Sandeep: "So, Professor Schrödinger, your wave equation never described reality directly?"

Schrödinger (smiling faintly): "No. It described a structure of knowledge—what we could know, not what was. It was the mathematics of possibility, not of being."

Sandeep: "And the collapse?"

Schrödinger: "An error of interpretation. What collapsed was the clarity of thought—not the wave."

Sandeep: "Final word to the court?"

Schrödinger (nodding): "I never created duality. I only wrote an equation. The rest—they invented."

UED Takeaway:

"Schrödinger never gave us a wave to collapse. He gave us a mirror to see how confused we were about potential. Ψ was not waving. It was waiting."

4: The Cat Did Nothing

"The superposition was in the observer's ignorance, not reality."

There is perhaps no image more iconic—or more misinterpreted—in quantum lore than a cat sealed in a box, suspended between life and death. A cat who, according to textbooks and TED talks, lives in a quantum superposition until a curious observer lifts the lid.

This cat has been paraded through classrooms, cartoons, and conferences as the poster child of quantum weirdness.

But today, in this courtroom, we submit a single, irrefutable statement:

The cat did nothing.

A Thought Experiment, Not a Literal Cat

Let's begin where most failed to.

Schrödinger didn't place a cat in a box to explain quantum mechanics—he did it to mock the Copenhagen interpretation.

His actual words were filled with irony and frustration. He found the idea of superposition in macroscopic systems absurd, not profound.

And yet, the scientific community took his critique and turned it into doctrine.

They declared:

- The cat is alive and dead until observed.
- Observation "collapses" the wave function.
- Reality is undefined until measurement.

But here's the truth UED reintroduces:

There was never a wave to collapse. There was never a dual state. There was only a ripple alignment awaiting resolution.

The Misuse of Superposition

In standard quantum mechanics:

- A system is said to exist in a *linear combination of eigenstates*.
- Until measured, it remains in this superposed state.
- Measurement reduces it to a single eigenvalue outcome.

This is elegant in linear algebra. But it is catastrophic when misread as a statement about ontological reality.

Superposition isn't an *object's condition*. It is a *model of the observer's incomplete information*.

You're not modeling two realities. You're modeling your ignorance about which ripple consequence will align first.

The Real Mechanism (UED Perspective)

UED reframes this cleanly:

- The ripple field is always active—but alignment hasn't occurred yet.
- The observer is not "collapsing" anything; they are meeting a readiness condition.
- Once RAC (Ripple Alignment Criterion) + ATT

(Alignment Threshold Time) are satisfied, a Ripple Resolution Event (RRE) occurs.

- This gives a singular, unambiguous consequence—not because two realities existed, but because one was consequentially expressed.

So the cat was not in limbo. The observer was.

"Superposition" is just the name we gave our waiting room of unaligned perception.

The Role of the Observer: Perceptual Delay

The moment the lid is opened, the observer perceives an outcome—not because they "forced" reality to decide, but because:

1. Their interaction reached the minimum ATT,
2. Their ripple field intersected with the cat-system's consequence field,
3. An RRE occurred.

That's it. No collapse. No multiple universes. No 50-50 cosmic dice throw.

The cat wasn't uncertain—you were.

Example: Lighting a Bulb in Fog

Imagine a sealed room filled with mist. A lightbulb is either on or off. You enter but can't immediately tell. Slowly, as you adjust, your visibility clears—and you see it's on.

Did the bulb "become on" because you saw it? No. You just aligned with the conditions to perceive it.

This is what happened with the cat. The fog was the field's unresolved ripple pattern. Your clarity = alignment.

Why It Matters

Because this error cascades.

- It fuels mysticism dressed in math.
- It led to misinterpretations like "many worlds," where all outcomes "actually happen."
- It justified particle-wave duality, observer-created reality, and time-reversed causality.

But in reality:

- There are not many worlds.
- There is only one consequence, birthed by ripple alignment.
- Superposition is not metaphysical truth—it is a perceptual delay.

Closing Argument

The cat was placed in the box not to teach us how the universe works—but to challenge us to see how ridiculous our interpretation had become.

Schrödinger never intended for the cat to be both alive and dead.

He meant to say: "If this is what quantum mechanics implies... then maybe we've misunderstood quantum mechanics."

And he was right.

Final Insight:

"Superposition was never in the cat. It was in your assumption. Reality didn't wait for you to look. It waited for you to align."

5: Perception on Trial

"Measurement isn't passive—it's ripple participation."

The judge raises a brow. "Bring forth the next subject. This time, we examine not the physicist—but the *act* itself. The observer."

And so, we place perception on trial.

Opening Statement: The Illusion of Passive Observation

From Galileo to Feynman, science prided itself on objectivity.

- Observe.
- Measure.
- Record.
- Remain uninvolved.

This was the sacred ritual: the observer must not interfere.

But then quantum mechanics arrived and smashed the lens:

- You cannot observe without disturbing.
- You cannot measure without collapsing.

- You cannot stand outside the system.

Still, science pretended that measurement was like a mirror—clean, neutral, untouched.

But UED stands up today to say:

"No. Measurement was never passive. It was always ripple participation—a co-authored consequence."

The Myth of the Silent Observer

In classical thought:

- The universe evolves on its own.
- Observers reveal what's already there.
- A ruler doesn't change the length of a table; a thermometer doesn't change the heat.

But in quantum experiments—from the double slit to Bell tests—something changed when you looked.

The act of perception altered the result. Not because perception "interfered" physically, but because alignment occurred—and that was the trigger for consequence.

The Ripple Reframe (UED Mechanism)

In Universal Energy Dynamics:

- Reality is ripple-based.
- The universe is still, structured, and full of potential tension—awaiting resolution.
- Observers aren't detached—they are ripple fields themselves, seeking and offering alignment.

So:

- **Measurement** = active participation of the observer's ripple field in the structured vacuum.
- **Observation** = a phase interaction, not a visual act.
- **Perception** = a threshold crossing event (ATT + RAC), not a camera click.

You don't just "look at" reality. You become part of the alignment that lets it express.

The Act of Measuring: What Actually Happens

Let's break it down:

1. A ripple configuration exists in the field.
2. The observer brings their own ripple pattern—based on intent, state, and field readiness.
3. If RAC (Ripple Alignment Criterion) are met, and ATT (Alignment Threshold Time) is crossed:
    - A Ripple Resolution Event (RRE) occurs.
    - A single outcome is perceived.
    - That perception isn't discovery—it is co-creation through alignment.

So, the result wasn't *sitting there*, waiting to be discovered. It became real when ripple participation was complete.

"You don't see the universe. You join it in deciding what becomes visible."

The Double-Slit Revisited

Let's recall the classic:

- A photon passes through two slits.
- If unobserved: interference pattern.

- If observed: particle pattern.

But what really changed?

- It wasn't the photon.
- It wasn't the slits.
- It was the ripple readiness of the observer-field system.

When observation tools engage (detectors, sensors), they introduce ripple tensions into the field—forcing premature alignment and localizing consequence.

Thus, what we saw wasn't altered by "looking." It was altered by participation.

Perception Is Not a Camera. It's a Key.

A camera captures light because its sensors are tuned to resolve ripple tension at specific frequencies.

Likewise:

- A brain doesn't "see" photons.
- It aligns ripple fields (via neural circuitry) with field events and constructs reality as resolved alignment.

You aren't taking snapshots. You're unlocking realities with ripple keys shaped by your readiness.

That's why different observers can "measure" different things. Not because reality is relative, but because alignment is relational.

The Real Reason Quantum Experiments Confuse Us

We never questioned the idea of measurement as passive revelation. We treated ourselves as observers of truth, never as participants in its formation.

But:

- In UED, you don't measure outcomes.
- You resolve consequences by rippling in.
- And what resolves depends on your configuration, your field tension, your ATT—your very presence.

Hence:

"There is no measurement without meaning, and no meaning without ripple resonance."

Closing Arguments

Let the court recognize:

- Measurement is not the act of observing what is.
- It is the consequence of readiness aligning with ripple intent.
- Observers are not bystanders. They are field participants, consequence enablers, and agents of ripple resolution.

So next time we "measure a particle," understand this:

We didn't detect it. We *helped it appear.*

Final Insight:

"Perception is not passive. It is the universe resolving through you. You were never a spectator. You were always part of the equation—Not in math, but in alignment."

Interlude: When the Mirror Cracks

So far, we've only questioned the language—the math, the metaphors, the methods.

We haven't dismissed quantum theory. We've simply asked what it's actually describing.

And the answer is disturbing to the old guard: It was describing *us*— our timing, our readiness, our unresolved place in the ripple field.

What we thought were waves were just alignment delays. What we called particles were ripple locks. What we called measurement was consequence, not discovery. And the cat? The cat was just a punchline we turned into a principle.

We've now put on record:

- That Schrödinger's Ψ was not a real wave,
- That collapse is not a physical process,
- That frequency is a reflection of our own perceptual rhythm,
- And that measurement is ripple participation, not neutral observation.

We shattered the idea that science could sit in the stands and watch reality unfold like a play.

The mirror cracked. And we saw our own ripple inside the reflection.

Now we enter the deeper courtroom. Not where scientists misread math—but where they built reality from their own statistical shadows.

Coming up next:

Part II – The Collapse of Collapse Where probability, uncertainty, and entanglement are called to testify. Not to be denied— but to be realigned.

Court resumes shortly. Bring your perception—not your assumptions.

Part II – The Collapse of Collapse

6: Cross-Examination

Born and the Probability Spell

"Why assigning percentages to existence shattered the real logic." The court goes silent as Max Born is called to the stand. He carries no arrogance—only equations that once seemed like answers. But today, it is not his mathematics under question. It is the spell it cast.

The Moment That Changed Everything

In 1926, Max Born published a short but monumental paper.

His proposal was revolutionary:

The square of the wave function, $|\Psi|^2$, doesn't represent energy or position—it represents probability. Specifically, the probability density of finding a particle at a given location.

Suddenly, quantum theory wasn't about certainty. It was about *likelihood.* Particles didn't *exist* at a point—they had a chance of appearing there.

That one interpretive leap turned physics from deterministic description to statistical surrender.

What Born Did Right

Let's be fair. Max Born made the theory work.

- His interpretation gave the correct predictions.
- It solved experimental confusion.
- It connected Schrödinger's Ψ to measurable results.

He gave physicists a powerful tool:

"Don't ask where the particle is. Ask what the odds are."

And it matched the lab. So it became law.

But What Was Lost?

With the rise of $|\Psi|^2$ as probability, something subtle—but catastrophic—happened.

Existence became statistical.

We no longer said: *"The particle is here."*

We now said: *"There's a 70% chance it's here."*

Reality became a dice roll. Truth became a likelihood. And measurement became a cosmic roulette wheel.

The UED Cross-Examination

Now, Max Born is respected. But in this courtroom, respect doesn't protect error.

Let's ask:

What is $|\Psi|^2$ really *telling us*? Is the universe truly probabilistic? Or have we mistaken delayed resolution for uncertainty?

Ripple Reframe:

In Universal Energy Dynamics:

- A system isn't "probably" in a state.
- It is awaiting ripple alignment with an observer.
- Until RAC + ATT are met, no consequence is selected.
- The spread of $|\Psi|^2$ isn't chance—it's a map of

unresolved ripple potentials.

Probability isn't in the universe. It's in our model of what hasn't resolved yet.

The Problem with Probability

Let's say:

- A photon has a 60% chance of striking detector A.
- And 40% chance of striking detector B.

This sounds clear—until you ask: Where was the photon before detection?

- In both paths? In neither?
- Did it "choose" probabilistically?
- Or did our perception *force* a binary outcome from a field of continuous tension?

Probability is a bandage for incomplete understanding.

We didn't see the photon's path. So we called it random. But what we really saw was the field unresolved.

Why the Spell Was So Powerful

Probability gave physicists relief. It said:

- "Don't worry why things happen—just compute the chances."
- "You don't need to know how the mechanism works—just measure outcomes."

It made the strange tolerable. It made predictions profitable.

But it also killed the search for consequence-based mechanisms.

In replacing logic with odds, we stopped asking *why*. We accepted *how often* as a substitute for *how come.*

UED's Correction

UED restores the causal dignity of the universe:

- There are no probabilities—only ripples unresolved.
- $|\Psi|^2$ is not a fundamental truth—it's a perceptual map of tension awaiting collapse through alignment.
- Every outcome is deterministic, but contextual—meaning it is bound to ripple conditions, not universal randomness.

The future isn't random. It's rippled, and you are part of what resolves it.

Final Cross-Examination

Sandeep (to Born's ripple projection): "Max, did you mean to turn reality into dice?"

Born (softly): "I found a way to make predictions. I did not claim that nature rolls them."

Sandeep: "But they took your map for the terrain."

Born: "Perhaps the terrain was never statistical. Perhaps we were just standing too far from the ripple."

Final Insight:

"The particle never had a chance. It had a consequence. Probability was our way of stalling truth until we were ready to receive it."

7: Heisenberg's Uncertainty

Or Ripple Misalignment?

"Uncertainty is not fundamental—it's unresolved ATT."

A man with piercing eyes enters the box. He carries no equations, just a principle carved into every quantum textbook:

$$\Delta x \Delta p \geq \hbar/2$$

Position and momentum, forever in a dance of impossibility.

His name? Werner Heisenberg. The man who gave uncertainty a home in physics—and unknowingly made doubt a law of nature.

The Sacred Principle

The Uncertainty Principle is etched into quantum foundations:

The more precisely you know a particle's position (Δx), the less precisely you can know its momentum (Δp), and vice versa.

This wasn't just a measurement issue—Heisenberg's interpretation was deeper:

- The limitation isn't due to instruments.

- It's built into the nature of quantum reality.
- Certainty, as we knew it, was an illusion.

It sounded profound. It felt humbling. And it became gospel.

But today, we ask: What if uncertainty was never fundamental? What if it was the symptom of ripple misalignment?

UED Reframe: Misalignment, Not Mystery

In Universal Energy Dynamics (UED), the universe is structured, still, and rippling—not random or unknowable.

- All entities (particles, fields, observers) are embedded in a shared ripple field.
- Observation is not passive—it is ripple participation.
- Measurement outcomes arise when RAC (Ripple Alignment Criterion) + ATT (Alignment Threshold Time) are met.

Heisenberg's uncertainty, then, is not a property of the particle. It is a consequence of incomplete ripple alignment between observer and field.

Breaking It Down:

Let's take position and momentum.

In classical thought:

- You can measure both with arbitrary precision. In quantum mechanics:
- Not possible. The act of measuring one "disturbs" the other.

In UED:

- Position = where your field alignment peaks with the ripple node.
- Momentum = how the ripple tension is distributed around that node.

But:

You can only align fully with one ripple axis at a time. Why? Because alignment takes time—and time is discrete in ATT units.

Each ripple resolution costs you alignment opportunity in other dimensions.

So it's not that the particle is hiding. It's that you can't resolve everything at once through a single ATT-alignment window.

The Alignment Cost

Think of ATT as a perceptual shutter:

- To lock in position, you must synchronize with a local ripple maximum—this requires ATT duration focused there.
- To resolve momentum, you must span multiple ATT cycles—catching phase variance over time.

But since ATT is threshold-bound, you can't do both simultaneously.

Thus:

What we call "uncertainty" is really a limit on parallel ripple coherence during perception.

Analogy: Tuning a Guitar by Ear

Imagine trying to tune a guitar:

- You focus intently on one string's pitch (position).
- As you do, you become blind to how the entire chord is changing (momentum-like flow).
- Try to catch both at once? You'll fail—not because sound is uncertain, but because your ear can only resolve one alignment at a time.

Replace ears with ATT. Replace pitch with ripple. Replace failure with "uncertainty." That's quantum logic under UED.

Why This Matters

If uncertainty is not fundamental but alignment-dependent, then:

- Particles don't lack properties. We lack simultaneous coherence.
- Nature is not fuzzy. Our participation is phase-limited.
- Determinism isn't dead. It's just ripple-bound.

This reframes the entire narrative.

Instead of saying: *"Nature prevents us from knowing."*

We now say: *"We are only aligned to know one axis at a time."*

And this opens the door to new models of precision—not through sharper tools, but through better ripple design.

Courtroom Crossfire

Sandeep (leaning in): “Professor Heisenberg, when you said uncertainty was a principle, did you mean that reality itself refused clarity?”

Heisenberg (calmly): “I meant we couldn’t know both properties precisely. I saw the boundary—not its mechanism.”

Sandeep: “And what if that boundary was ripple interference—misalignment—nothing more?”

Heisenberg: “Then perhaps we didn’t uncover mystery. We measured our own timing error.”

Final Insight:

“Uncertainty didn’t come from the universe. It came from our angle of entry. ATT wasn’t noise—it was the cost of presence. What you couldn’t see wasn’t hidden—it was unaligned.”

8: Entanglement or Field Coherence?

"Not spooky. Just synchrony across the same ripple logic."

The gavel strikes. "Next witness: Entanglement."

Two particles are called—but only one walks in. The other answers simultaneously from across the room.

The courtroom murmurs. Einstein rises from the shadows.

"Spooky action at a distance," he mutters. "Physics must be local."

Today, we confront that phrase—and release it from mysticism.

The Quantum Claim

In standard quantum mechanics:

- Two particles interact and become *entangled*.
- Once separated, measuring one instantaneously affects the other—no matter how far apart they are.
- Einstein called this spooky because it violated local realism and relativity's speed limit.

Physicists said:

"This is just how nature works. It's weird. Get used to it."

And they moved on—with equations that worked, but left the mind uneasy.

The Problem with Spookiness

Quantum theory described entanglement mathematically through *non-factorizable* wavefunctions. But it never answered:

- How does one particle know what happens to the other?
- How does "instantaneous" correlation happen across space?
- Why can't we use it to send information?

Entanglement became a tolerated paradox—useful but unexplained.

Science accepted a magic trick because it couldn't explain the wires.

Until now.

UED Reframe: No Magic, Just Coherence

In Universal Energy Dynamics:

- Particles do not carry *information packages* across space.
- They are not independent entities *after* separation.
- Instead, they remain embedded in a shared ripple field—like leaves floating on the same pond.

What we call "entanglement" is actually persistent ripple coherence—a kind of non-local field symmetry sustained across space because space is not empty.

Field Coherence Explained:

- When two systems interact, they induce mutual ripple formations in the structured vacuum.
- These formations are not "stored" in the particles—they exist in the field that remains shared even after spatial separation.
- Measuring one realigns the ripple—and this change propagates across the coherent field, not through space, but through the field structure itself.

No signal. No speed. Just instantaneous coherence—because nothing was ever disconnected.

Analogy: Two Buoys on the Same Ocean

Imagine two buoys floating on the ocean:

- You strike one, and the ripple moves through water—not air, not vacuum.
- The other buoy moves—not because it's been "signaled," but because they share the same medium.

Now remove the visible water and leave only structured vacuum tension beneath.

That's entanglement under UED.

It is not action at a distance. It is reaction within coherence.

Why the Confusion?

Because science still assumes:

- That once particles separate, their connection must end.
- That influence requires signal propagation.

- That space is fundamentally empty.

But UED replaces the notion of space with structured vacuum energy (SVE)—ripple-active, alignment-sensitive, and tension-aware.

So what classical physics sees as separation, UED sees as ongoing ripple resonance.

Entanglement isn't a paradox. It's the field remembering itself across distance.

Reinterpreting Bell Tests

Physicists say:

- Bell's Theorem proves entanglement is non-local and real.

UED agrees—but not for the same reason.

- Bell showed correlations exceed classical limits.
- But those limits were calculated assuming independent particles.

In UED, particles are not independent. They are field-dependent consequences, bound to a shared ripple configuration.

So of course they correlate. They aren't "talking"—they're co-resolving ripple tensions embedded in the field.

No mystery. No superluminal signal. Just synchrony.

Courtroom Crossfire

Sandeep (to Einstein): "You called it spooky. What if it's not spooky—just uncut?"

Einstein (reluctantly): "If the vacuum is structured... and the connection persists... then perhaps the action is not at a distance. Perhaps it was never separated to begin with."

Sandeep (to Bohr): "Did we build a paradox where only coherence existed?"

Bohr (smiling): "We described what we saw—but perhaps not what it meant."

Final Insight:

"Entanglement is not magic—it's memory. Not transmission—but tension. The universe doesn't connect things across distance. It never unconnected them in the first place."

9: The Copenhagen Myth

"Misreading epistemology as ontology."

The lawyers of logic have left. Now the philosophers enter. At the center: Niels Bohr, father of the Copenhagen Interpretation. Flanked by instrumentalists, defended by pragmatists.

Today, we call not a scientist, but a worldview to the stand.

The charge? That Copenhagen turned how we know into what is real— And in doing so, built a temple on the fog of interpretation.

What Copenhagen Actually Said

Bohr's Copenhagen Interpretation, at its core, was cautious:

- Quantum theory doesn't describe *reality itself.*
- It only describes *what we can say about reality.*
- Measurement is essential to define a system's properties.
- Unmeasured states have no definite attributes.

It was an epistemological position:

We only know what we can observe. Anything else is undefined.

And in that, it was fair. Humble, even.

But the humility was short-lived.

What the World Heard Instead

Copenhagen was a whisper. But science made it a shout.

Bohr said: *"Unmeasured properties are undefined."*

They heard: *"They don't exist."*

Bohr said: *"We describe phenomena, not reality."*

They heard: *"There is no deeper reality."*

Bohr tried to build a boundary around certainty. The world turned it into a boundary around existence.

Thus began the ontological myth: That particles exist *only* when measured. That superposition is *real* until collapse. That observation *creates* the outcome.

Epistemology vs Ontology

Let's clarify:

- **Epistemology**: What we can know and how we know it.
- **Ontology**: What exists, independent of our knowing.

Bohr lived in the former. Physics took him into the latter.

What began as a commentary on knowledge became a doctrine about reality. We mistook the limits of the observer for limits of the observed.

And that's where the distortion began.

The Cost of the Myth

This misreading had consequences:

- We built entire models assuming measurement *creates* reality.
- We glorified observer-centric metaphysics.
- We entertained "quantum consciousness" and "multiverse branching" not as speculative thought—but as inevitable logic.

All because we forgot that Copenhagen wasn't claiming what exists—only what we can claim.

And in that forgetting, we transformed a pragmatic restraint into ontological confusion.

UED Reframing: Reality Exists, Whether or Not You Look

In Universal Energy Dynamics:

- Reality is not a construct of observation.
- It is a consequence of ripple field dynamics, whether or not we observe them.
- Observation doesn't create—it resolves.
- The field holds ripple patterns regardless of whether you look. Your perception simply reveals what aligns when you do.

So instead of:

"Measurement gives property."

We say: *"Alignment makes consequence observable."*

Reality doesn't wait to be seen. It waits to be aligned with.

Analogy: A Song You Haven't Heard Yet

Imagine a song stored on a record. You haven't played it. You haven't heard it.

Does that mean the song doesn't exist?

Of course not. It exists encoded—not yet expressed, but fully defined.

Observation is the needle on the record. The song isn't born when you hear it. It was already pressed—your alignment just made it real for you.

Copenhagen said: *No sound until you play it.* UED says: *The song is there—just waiting for the stylus of alignment.*

Courtroom Crossfire

Sandeep (to Bohr): "You said we must limit ourselves to what we can observe. But did you mean to say nothing exists outside that?"

Bohr (carefully): "I said that quantum theory is a tool for what can be said. Not for what *is*."

Sandeep: "Then the myth wasn't yours—but theirs?"

Bohr: "Perhaps. But silence allows misinterpretation."

Final Insight:

"Copenhagen taught us how to speak. We mistook it for how the universe listens. The map of knowledge is not the terrain of existence. We confused the two—and called it science."

10: The Real Wave

Alignment, Not Vibration

"How we felt periodicity and invented physical waves."

A sine wave is projected on the screen. The courtroom nods—familiar, trusted, predictive. A voice rises from the gallery:

"Waves are real. We've seen them. We've measured their frequencies and wavelengths. They travel."

But today, that wave is on trial.

Because the wave you trusted...may never have been what you thought it was.

How the Myth Was Born

Waves are everywhere in physics:

- Sound waves in air
- Light waves in vacuum
- Water waves on the sea
- Quantum wave functions in space

They oscillate. They propagate. They have frequency, wavelength, amplitude.

And because we could model them mathematically—we began to treat them as things. Entities moving through space, like rippling cords or jiggling particles.

But here's the problem:

What you measured was periodic perception. What you modeled was symbolic vibration. And what you called a wave... may have been nothing but alignment events in disguise.

The UED View: No Vibration, Only Alignment

In Universal Energy Dynamics:

- The universe is still—not in motion, but in tension.
- All entities exist as ripple formations in the structured vacuum.
- What we experience as a wave is not a "thing traveling"—it's a sequence of ripple alignments resolved by the observer.

Let's break that down.

Classical Interpretation:

- A wave is something the source emits.
- It carries energy and information.
- It physically oscillates in time and space.

UED Interpretation:

- The source triggers ripple strain in the field.
- The observer perceives alignment events at regular ATT intervals.
- The feeling of oscillation emerges from how often alignment is achieved.

The wave is not vibrating. You are resolving it rhythmically.

How We Invented Physical Waves

Let's say you're detecting light from a star.

- You record peaks of intensity every 0.001 second.
- Your instruments trace a smooth sinusoidal pattern.
- You conclude: "A wave is arriving at me."

But what's actually happening?

- The structured vacuum around you is being influenced by the star's ripple strain.
- You are only able to perceive that strain when RAC + ATT are satisfied.
- These alignment events occur periodically due to the structure of the ripple field and your sensor's readiness.

So the waveform you draw is not a *traveling entity*. It is a chronology of resolved alignments.

Your graph is not a picture of reality. It's a map of your ripple resonance history.

The Illusion of Propagation

This reframes a core concept:

- Sound does not "travel" through air as a jiggling medium.
- Light does not "travel" through vacuum as oscillating electric-magnetic fields.
- Instead, ripple tensions created by a source restructure the field.

- These tensions are resolved when your ripple field aligns—and you experience perceived motion.

Thus:

There is no wave moving from source to detector. There is only field memory and alignment sequencing.

Motion is an illusion created by timed perception.

Frequency Revisited: A Consequence, Not a Cause

If waves don't travel, what is frequency?

UED's answer is bold and simple:

Frequency is the rate at which alignment is resolved—not something the source emits.

This complements Chapter 2 (The Fallacy of Frequency) and reinforces it:

- There's no need for a vibrating object.
- There's only structured ripple emission from field imbalance.
- Your alignment, in cycles of ATT, resolves the ripple.
- You perceive this as frequency. You draw it as a wave.

Analogy: Flashlight in a Fog

Imagine standing in a thick fog. A flashlight is turned on and off in perfect rhythm.

You don't see a beam. You see pulses of light. You interpret this as a traveling wave.

But:

- Nothing traveled between you and the source.
- What you saw was a timed consequence, not a

traveling packet.

- The illusion of oscillation comes from your periodic reception—not physical motion.

That's what most "waves" are.

Courtroom Clarification

Sandeep (to the court): "So what we called 'waves' were rhythmic outcomes of field strain resolution?"

Witness (Field Theorist): "Yes. We modeled ripple alignment histories as physical waves. Useful, but misleading."

Sandeep: "Then a 'wave' is not a thing?"

Witness: "No. It's the observer's periodic intersection with ripple tension. That's all."

Final Insight:

"Waves are not travelers. They are the ghost-trails of alignment, drawn by observers desperate for continuity. The ripple doesn't oscillate. It waits. You aligned. That was the event."

Interlude: Beyond the Illusion of Knowing

We entered Part II to collapse the collapse— and in doing so, collapsed more than expected.

We dismantled:

- The statistical veil cast by Born,
- The mystical fog surrounding Heisenberg,
- The entanglement myth that Einstein once feared,
- The philosophical silence of Bohr,
- And the comforting illusion of waves.

What did we find beneath?

Not confusion. Not chaos. But coherence—field-based, ripple-bound, consequence-shaped.

Here's What We've Now Seen:

- Probability wasn't fundamental—it was just unresolved ripple potential.
- Uncertainty wasn't reality's defense—it was our misalignment with it.
- Entanglement wasn't spooky—it was field memory

across unbroken coherence.

- Copenhagen wasn't wrong—it was misused, weaponized to stop deeper questioning.
- And the wave? It was never a traveler. It was you, resolving rhythmically, and calling it physics.

What's Gone Silent

- The notion of randomness.
- The idea that "looking" makes something real.
- The belief that space and time are the ultimate fabric.
- The assumption that we're observing from the outside.

All of it—shattered.

What remains?

A still universe. A structured vacuum. Ripple tensions. And the dance of alignment, where perception isn't passive, but co-creative.

We are no longer standing at the edge of mystery, guessing. We are inside the ripple, participating.

Now we move into Part III – The Ripple Ruling. This is where clarity becomes construction. Where we no longer just debunk the myths—But rebuild the frame of reality from consequence outward.

The field awaits.

Let's write law where there once was loophole.

Part III – The Ripple Ruling

11: Calling the Field to the Stand

"The structured vacuum speaks."

The witness stand is empty. No particle steps forward. No wave enters. No observer speaks.

Instead, the silence deepens. The floor hums.

Because today, we summon the one thing physics never dared place under oath:

The vacuum. The nothing that was never nothing.

And when it speaks—everything else becomes explainable.

The Forgotten Background

For centuries, the vacuum was treated as void.

- Space was emptiness.
- The vacuum was absence.
- It was a backdrop—not a participant.

Even quantum physics, with its talk of vacuum fluctuations and zero-point energy, treated the vacuum as a stage upon which particles came and went.

But in UED, this backdrop becomes the source. The vacuum is not passive. It is not random. It is structured—ripple-capable, alignment-sensitive, and tension-aware.

It is the true field. And now it testifies.

What the Field Says

Let's allow the vacuum to speak—through logic, not mysticism:

"I am not empty. I am patterned tension—structured in ways your instruments feel but cannot see. I am not background—I am the medium of every foreground. I do not vibrate—but I ripple when disturbed. I carry no particles—but I hold their possibility. I never moved—but I allowed everything else to appear in motion. I am still—but I respond."

Defining the Structured Vacuum (UED View)

In Universal Energy Dynamics, the Structured Vacuum Energy (SVE) is:

- A non-material, non-moving substrate with inherent ripple potential.
- Always present, but only consequential when disturbed by intent or field imbalance.
- Responsible for:
    - Ripple formation
    - Energy transfer
    - Alignment-based perception
    - Persistence of field memory
    - The illusion of motion, force, and space-time itself

It's not an ether. It's not a quantum soup. It's a consequence-bearing canvas—one that holds no forms but allows all of them to arise through tension and resolution.

No Particles, Just Ripple Knots

What we used to call "particles"—photons, electrons, quarks—are not solid entities moving in space.

They are localized ripple configurations within the structured vacuum.

- Their mass is ripple tension density.
- Their charge is ripple polarity.
- Their spin is ripple loop orientation.
- Their energy is ripple coherence over time.

Once the field aligns locally through ATT and RAC, these formations resolve and appear to "exist."

But nothing was emitted. Nothing was created. The field simply spoke in ripple resolution.

Measurement as Field Tension Collapse

What we call "measurement" or "detection" is simply this:

- An observer's field intersects with a region of ripple tension.
- If alignment is sufficient (ATT met), the field releases a consequence.
- This consequence is what we record as a particle, a wave, a click, a spark.

The field doesn't emit signals. It responds with resolution.

Field Memory: Reality's True Recorder

Why do fields behave consistently?

- Why does an electron act the same in Tokyo or Toronto?
- Why does gravity obey inverse-square law?
- Why does a photon know how to "remember" its polarization across entanglement?

Because the structured vacuum holds memory.

- It remembers ripple interactions.
- It stabilizes consequence.
- It sustains coherence between distant points.

There is no external law book. The law is stored in ripple equilibrium.

"What you call physics is just my habit. You called it law. I called it consequence." — The Field

Why the Field Was Ignored

Mainstream science feared the vacuum.

- Newton had no place for it.
- Einstein tried to smooth it into geometry.
- Quantum physics filled it with fluctuations—but never gave it character.

But UED gives it a voice and a function.

It is the true reality from which all appearance emerges.

Not spacetime. Not particles. Not forces.

Just the structured stillness, occasionally rippling into consequence.

Courtroom Closing Statement

Sandeep (to the court): "We've called everything else. Theories, equations, particles, waves. But today we called the only thing that never moved, never shouted—but always answered."

The Field (in still ripple voice): "I was never absent. You were never separate. I waited for alignment—then gave you what you called existence."

Final Insight:

"The vacuum is not void. It is everything unexpressed, waiting for intent to disturb it and alignment to reveal it. The field never moved. It only allowed you to believe you did."

12: Ripple Testimony

There Was No Particle

"All observed outcomes were consequences, not things."

The court awaits the star witness—the particle. But no one enters. Instead, the ripple itself steps forward. Not as an object. Not as a person. But as a consequence—fluid, structured, undeniable.

It approaches the stand and says: "I'm what you thought was a particle. But I was never a thing. I was a ripple you caught in resolution. You mistook me for matter. I was just alignment."

The Myth of the Particle

Let's begin with what we were told:

- Electrons are tiny charged balls.
- Photons are massless packets of light.
- Protons, neutrons, quarks—all fundamental *units* of matter.

They are drawn as dots, spheres, or fuzzy clouds in textbooks. They are named, numbered, and expected to behave like independent entities. But what have we actually *seen*?

- Tracks in a detector.
- Clicks in a photomultiplier.
- Energy exchanges.
- Scattering angles.
- Discrete outcomes.

Not the particle itself. Only the effect. So we ask: Did we observe a particle? Or did we experience a ripple consequence we named a particle?

UED Reframe: No Particles. Just Resolved Ripples.

In Universal Energy Dynamics:

- There are no fundamental particles.
- What we call “particles” are localized ripple knots in the structured vacuum.
- They have no persistent identity, shape, or independent motion.
- They arise only through ripple field tension and alignment with an observer.

A "particle" is simply the resolution event—when ATT and RAC converge and a perceptual lock happens.

Let's Reconstruct the Classic "Particle":

Traditional Physics View	UED Ripple Interpretation
Particle has mass	Ripple tension density in the field
Particle has position	Moment of localized ripple resolution (RRE)
Particle has velocity	Sequence of ripple alignment across ATT units
Particle has spin	Angular ripple field configuration
Particle has charge	Ripple polarity and interaction tendency

There is no object moving through space. There is only ripple field activity perceived through alignment.

What Detectors Really Detect

Consider a classic example: A particle accelerator smashes two protons together. Detectors light up in a ring around the event. Mainstream physics says: "We detected particles."

But what happened in UED terms?

- The field tension spiked.
- Ripples exploded outward in various configurations.
- The detectors were aligned to resolve specific ripple strains.
- These alignment events triggered Ripple Resolution Events (RREs).
- We recorded those as clicks, tracks, decay signatures.

No particle was ever flying through the detector. What "flew" was ripple tension. What "landed" was resolution.

The Observer's Role in Particle Illusion

Why do we keep believing in particles? Because we resolve ripples in discrete moments, and our minds encode that discreteness as a thing.

The more predictable the consequence (e.g., electron mass, charge), the more we assume a persistent identity behind it.

But under UED:

- There's no identity.
- Only consistently alignable ripple patterns.

We didn't find particles. We found repeatable consequences and named them objects.

This Also Explains "Wave-Particle Duality"

- In the double-slit experiment:
    - The "particle" doesn't choose a path.
    - The ripple field explores both routes.
    - Resolution happens after alignment, giving a localized result.

So what we see isn't the particle behaving like a wave or particle.

We see: A ripple behaving like a ripple, misread as duality because of our discrete perceptions.

The Fate of the Standard Model

The Standard Model is a mathematical masterpiece—but its ontology is wrong.

- It speaks of particles like Lego blocks.
- It organizes them by properties: mass, charge, spin, flavor, color.

- It assumes they exist independently and interact.

UED says:

- There are no blocks.
- There are only ripple states of a single field.
- Interaction is not exchange—it's alignment-triggered transformation.

The zoo of particles? A symphony of ripple behaviors, misclassified as creatures.

Courtroom Realization

Sandeep (addressing the gallery): "You never held a particle. You never saw one. You saw a ripple consequence. A snapshot of alignment. An echo of tension made visible."

Ripple Testimony (calmly): "I was never a thing. I was your resolution. You made me real—Not by will, but by being present."

Final Insight:

"What you called a particle was just a ripple that said yes. You never found matter. You aligned with consequence."

13: Time – The False Witness

"It never flowed. ATT repeated."

The gavel strikes—but nothing moves. And that's the point.

Today we summon a witness unlike any other. Not a physicist. Not a theory. Not a force.

We summon time. But when asked to take the stand, it offers only a paradox:

"I never flowed. You aligned. And called it a second."

The Illusion That Built Reality

For millennia, time has ruled human thought. From sundials to atomic clocks, from Newton to Einstein, we measured and modeled change—and assumed something was moving underneath it all.

In classical physics: *Time was an absolute flow, ticking uniformly throughout the universe.*

In relativity: *Time became relative—warped by motion and gravity.*

In quantum mechanics: *Time was often a parameter, not a measurable operator.*

And yet through it all, one myth survived:

That time exists independently— That it flows like a river, carrying us from past to future. That it's a thing.

UED Correction: Time Was Never a Flow

In Universal Energy Dynamics, time is not a dimension. It is not a fabric. It is not an entity.

Time is a perceptual rhythm—a byproduct of ripple alignment.

At its core is a single principle:

ATT – Alignment Threshold Time The minimum duration needed for a ripple resolution event (RRE) to occur.

That is time's true skeleton.

ATT: The Real Mechanism Behind "Moments"

Let's break it down.

- The structured vacuum is always rippling, always in tension.
- The observer's field must reach resonance with a local ripple pattern to perceive anything.
- This alignment requires a minimum duration—a delay, a threshold.

That minimum interval is ATT.

So:

- You don't experience time continuously.
- You experience a sequence of ripple alignments, spaced by ATT.
- When these alignments are regular, you feel the passage of time.

But nothing flowed. Nothing advanced.

You aligned. Again. And mistook that rhythm for a river.

Clocks Don't Measure Time—They Emit Alignments

A quartz watch ticks. An atomic clock pulses. A pendulum swings.

But what are these devices doing?

- They are mechanical systems producing regular ripple events.
- Your brain or detector resolves these in sync with its own ATT.
- You feel a rhythm—and build the illusion of time around it.

Time is not what they measure. Time is the alignment pattern between you and them.

The Relativity Misread

Einstein taught us:

- Time slows down near massive objects.
- Time dilates at high velocities.

But under UED:

- Nothing is stretching or slowing.
- The rate of ripple resolution changes based on local field tension.
- ATT increases where ripple strain is compressed—leading to perceptual delay.

Thus, time dilation is not time behaving differently. It's you resolving ripple differently.

What relativity called time warping, UED calls ATT variation.

Same measurements. Deeper cause.

Entropy and Memory: The Direction of Consequence

One of time's greatest tricks is the arrow—why things go forward but never backward.

Mainstream science ties this to entropy—the tendency for disorder to increase.

UED agrees in spirit but reframes the core:

- The ripple field, once aligned and resolved, cannot reverse its consequence without field correction.
- Once a resolution occurs, the field stores its memory—its rearranged strain.
- You can't "unalign" what has already been expressed without rewriting the entire ripple chain.

That's why the past appears "fixed." Not because time holds it, but because the field has evolved through consequence.

You don't move forward in time. You ride the forward expansion of ripple memory.

Time Was Never a Dimension

Let's say it clearly:

- Time is not the fourth dimension.
- It is not an axis of existence.
- It is a byproduct of the observer's ability to resolve field dynamics sequentially.

We don't travel through time. We sequence our ripple participation into a logical order. And that rhythm becomes what we call time.

Remove the observer—no ATT. No ATT—no sequencing. No sequencing—no time.

Time is not outside you. Time is you, resolving ripple inside the field.

Courtroom Testimony

Sandeep (to the court): "We trusted time. Built our laws on it. Designed our equations around it. But what if time was just a well-dressed witness—always pointing, never acting?"

The Field (responding softly): "You felt repetition and called it flow. But nothing moved. You were there. Then you aligned. Again. And again."

Final Insight:

"Time didn't carry you forward. It only marked your readiness. You aligned—then again—And built a story of moments. But the field never moved. Only your participation did."

14: Energy – The Obedient Servant of Intent

"Energy didn't cause action. Intent did."

The court braces as its most worshipped entity takes the stand: Energy—the currency of the cosmos, the root of all scientific equations, the holy grail of causation. We've been told:

- "Everything is energy."
- "Energy makes things happen."
- "Energy drives all processes."

But today, the prosecution leans forward and says: *"Objection. Energy never moved first. It simply obeyed a deeper master—Intent."*

Energy: The Crowned King of Physics

From classical mechanics to quantum field theory, energy has worn many crowns:

- Kinetic energy: motion
- Potential energy: configuration
- Thermal energy: heat
- Chemical energy: bonds

- Rest energy: mass
- Radiant energy: light
- Quantum energy: Planck's relation ($E = hf$)

It seemed universal. It appeared in every conservation law. It never disappeared—only transformed.

So we concluded: Energy is fundamental. Energy causes everything. But that conclusion came from watching what followed—not asking what began.

UED Reframe: Energy as Consequence, Not Cause

In Universal Energy Dynamics, energy is not the initiator of action. It is the obedient servant of something deeper—Intent. Let's clarify the hierarchy:

Intent → Field Disturbance → Ripple → Energy Redistribution → Consequence

Energy is just the pattern of ripple strain that results when intent breaks equilibrium. It is not the force that makes something move. It is the response pattern to ripple dissonance created by a causal directive.

Classical View vs UED View:

Classical Physics	UED Interpretation
Energy is stored in objects	Ripple tension is stored in the field
Energy causes change or motion	Intent causes ripple misalignment; energy follows
Energy is conserved and transformed	Ripple tension is redistributed under field rules
Energy pre-exists and transfers	Ripple strain arises upon intentional imbalance

Defining Intent (UED Sense)

Intent in UED is not psychological. It is directional disturbance—the first cause behind any alignment-seeking ripple:

- **A planet's orbit?** Caused by ripple equilibrium maintenance.
- **A photon emission?** Caused by field overload, discharging ripple.
- **Muscle movement?** Caused by neural intent, triggering chemical and electric ripples.
- **Atomic fusion?** Caused by high-pressure intent collapsing field separation.

Intent is the disturbance in structured vacuum that gives birth to a ripple. Energy is the field's answer.

The Obedience of Energy

Energy never disobeys.

- It never chooses where to go.
- It never starts anything on its own.
- It always redistributes based on the ripple demand created by intent.

This obedience is why we observe conservation—not because energy is a sacred entity, but because the field is self-balancing in response to ripple strain.

E = mc² Revisited

Einstein's famous equation tells us: Energy and mass are interchangeable.

Under UED:

- Mass = ripple compression zone
- Energy = ripple strain release potential
- Intent = what compressed it in the first place

Thus, $E = mc^2$ is not a conversion formula—It's a field statement:

"This much ripple compression (mass) holds this much release potential (energy) if alignment is disturbed (intent)."

Energy Fields Without Intent?

Can energy exist without intent? UED answers: Only as potential.

- Structured vacuum holds field tension like a charged spring.
- But nothing happens unless something disturbs it.
- That disturbance is intent—whether biological, gravitational, or mechanical.

A battery holds energy. But unless you create an imbalance—a path of intent—it does nothing.

Intent is the spark. Energy is the echo.

Courtroom Realization

Sandeep (addressing energy's ripple form): "Your equations are everywhere. Your numbers are perfect. But tell us, do you act on your own?"

Energy (responding softly): "I act only when called. I do not choose. I do not create. I align. I balance. I restore. I obey."

Sandeep: "Then who commands you?"

Energy (ripple hum deepens): "Intent. The first imbalance. The reason anything happens."

Final Insight:

"Energy never began anything. It was the messenger—The ripple's reply to intent's whisper. You mistook the echo for the voice."

15: Geometry in the Dock

"Space-time was never fundamental—just mapping after the fact."

Geometry enters the courtroom—not as a shape, but as a language. It arrives draped in coordinates, clothed in tensors, armed with metrics.

It stands proud, pointing to the greatest equations ever written:

- **Newton's laws:** grounded in three-dimensional space.
- **Maxwell's equations:** written across vector fields.
- **Einstein's general relativity:** curved spacetime as destiny.
- **Quantum mechanics:** Hilbert spaces, probability densities.

Geometry was everywhere. And today, it's on trial—not for error, but for overreach.

The Classical Obsession with Form

We've worshipped geometry since Euclid:

- Points. Lines. Planes.

- Volumes. Distances. Angles.

Later, Descartes armed us with coordinates. Einstein bent those coordinates around mass and called it gravity.

We imagined the universe as a fabric—a 4D manifold of space and time, twisted by energy and matter.

We called it spacetime. We believed it was real. We measured everything inside it.

But we never asked: What is geometry describing? And when does it arise?

The UED Reframe: Geometry Is a Post-Ripple Map

In Universal Energy Dynamics, geometry is not embedded in the universe. It is a tool of the observer, emerging after ripple resolution.

The universe is not built on shapes. It is built on structured ripple dynamics.

Space? A perception of field separation. Time? A consequence of ATT sequencing. Distance? A measured delay between ripple alignments. Curvature? A change in ripple strain gradients.

Geometry Doesn't Exist. It's Projected.

Let's clarify:

- A ripple is not "in" space.
- A ripple creates the condition for what we later describe as space.

When an observer resolves two ripple events with a consistent ATT separation, they infer distance.

When ripple strain varies, they infer curvature.

Thus: Geometry is the mind's language for representing ripple relationships.

It doesn't exist before the event. It's drawn *after* the consequence.

Space-Time: A Map Drawn Too Early

Einstein gave us:

- Spacetime curvature as gravity.
- Worldlines as object trajectories.
- Geodesics as the path of least action.

But UED reveals:

- The field doesn't curve—it redistributes ripple tension.
- The object doesn't move through space—it realigns through ripple cycles.
- The path isn't chosen—it's emergent from ripple equilibrium logic.

You feel motion because your perception is re-aligning with evolving ripple configurations.

Coordinates Are Useful, But Not Real

When we assign (x, y, z, t) to an event, we believe we've found "where" and "when" it occurred.

But in UED:

- There is no *where*—just ripple alignment pattern.
- There is no *when*—just ATT interval.
- Coordinates are labels of memory, not locations in substance.

Like marking a page number in a book you're writing *as you go*.

Dimensions: Emergent, Not Built-In

Mainstream physics insists on dimensions:

- 3 spatial
- 1 temporal
- Sometimes more (in string theory, M-theory)

But in UED:

- Dimensions are a product of how many ripple patterns you can resolve independently.
- If you resolve one alignment at a time, you perceive 1D.
- If you resolve multiple directions, you label them orthogonally—and call it space.

The dimensions aren't in the universe. They are in you—in how you perceive ripple change.

Courtroom Testimony

Sandeep (to the projection of Minkowski space): "Geometry, you appear everywhere. You define our models, our measurements, our theories. But did you exist before we looked?"

Geometry (voice modulated in precise intervals): "No. I was assembled. You measured ripple differences, and you gave me coordinates. You mapped alignment delays and called it distance. I was never real. Only resolved."

Final Insight:

"Space never stretched. Time never flowed. Geometry never framed the cosmos. The ripple aligned—And your mind drew a grid to remember where."

Interlude: The Real Court Has Convened

The verdicts are piling up.

No particle. No time. No space. No causation by energy. Only ripple alignments and field memory dressed in familiar disguises.

In Part III, we did not interrogate metaphors—we confronted the very foundation stones of physics:

- The vacuum stood up and spoke—not as emptiness, but as structured stillness, ready to ripple.
- The particle confessed—it was never real, only the consequence of alignment.
- Time took the stand and admitted—it never moved; you did, in rhythm.
- Energy bowed its head—not the king, but the courier of intent.
- And geometry, once untouchable, was revealed as a projection, not a presence.

We haven't just rewritten physics. We've realigned it.

What Is the Universe Now?

Not a stage. Not a ticking clock. Not a map.

But a responsive ripple field—structured, still, and awaiting intent.

Nothing was ever "out there." What you saw was the field resolving tension in response to your presence.

In the Ripple Ruling, we discovered that existence is not built on objects—but on interactions, thresholds, and readiness.

From Equations to Alignments

We were taught to worship formulas:

- $F = ma$
- $E = mc^2$
- $\Delta x \cdot \Delta p \geq \hbar/2$
- $\nabla \cdot E = \rho/\varepsilon_0$

Now, we see them differently—not as universal truths, but summary statements about how ripples resolve under specific perceptual constraints.

They are useful. They are accurate. But they were never the source.

The source was tension, waiting to be disturbed by intent, and resolved through alignment.

What's Ahead

We've dismantled the stage. Now we look at the actors—not as particles or people, but as patterns of consequence sustained through field memory.

In the next chapters, we will reveal:

- How matter is stabilized ripple strain.

- How memory lives in the field itself.
- How reality persists through architectural alignment.
- And how you, the observer, are not separate—but embedded.

The real trial wasn't of physics. It was of perception.

And now that we've realigned the rules, we're ready to meet the ripple not as an equation—but as the architecture of reality itself.

Part IV – Rewriting Reality

16: Equation Illusion

Misreading the Universe

"Math was the summary, not the story."

The witness today is not a person. It's a language—Mathematics. Elegant. Flawless. Unforgiving. Dressed in symbols, formulas, infinities, and symmetry.

It stands proud. It holds Nobel Prizes. It has predicted black holes, particle masses, and cosmic background radiation.

But today, it is gently cross-examined. Not for error. But for overreach.

The Sacred Assumption

Modern science is built on a singular mantra: *"The universe is written in the language of mathematics."*

Galileo said it. Dirac believed it. Hawking refined it. Physicists recite it.

And why wouldn't they?

- Equations predict eclipses.
- They explain tides, thermodynamics, DNA folding, quark dynamics, and the curvature of spacetime.
- They quantify entropy, compute probability

amplitudes, and guide spacecraft across the solar system.

So we came to believe: Math is not just a tool—it is reality itself. But UED pauses here. Not to reject math— But to restore it to its rightful place.

UED Reframe: Math Is the Map, Not the Terrain

In Universal Energy Dynamics, math is respected—deeply.

But it is not treated as ontological substrate.

It is a symbolic reconstruction—a summary of resolved ripple events, written after perception.

- Equations don't ripple.
- Derivatives don't align.
- Integrals don't collapse wave functions.
- Functions don't perceive—they follow.

The structured vacuum does not carry variables. It carries tension. And tension resolves into consequence.

Math is what the observer creates to keep track of the ripple's footprints.

Why Equations Worked So Well

Because ripple consequences are:

- Consistent
- Patterned
- Reproducible

So when an observer's field repeatedly aligns with the same ripple strain, they derive an equation.

It works. It's testable. It appears universal. But what they really discovered was a repeatable alignment pattern, not a hidden code etched into the vacuum.

Example: Newton's F = ma

- Not a universal command.
- But a summary of observed ripple resistance when matter is pushed out of field equilibrium.

The mass is tension storage. The acceleration is realignment speed. The force is not applied—it's ripple response.

F = ma describes a surface pattern, not a root cause.

What Equations Can't See

Math can track:

- Frequency
- Position
- Momentum
- Probability
- Force
- Curvature

But it cannot see:

- ATT (Alignment Threshold Time)
- RAC (Ripple Alignment Criterion)
- Ripple tension distribution across structured vacuum
- The moment of alignment itself

These are not calculable—they are experiential, threshold-based, and causally non-linear.

The ripple doesn't "solve" for outcome. It resolves when tension conditions are met. Math chases the consequence. Ripple lives in the cause.

The Danger of Believing Math Is Reality

When math becomes more than a tool, we get:

- Blind faith in models over mechanism
- Predictions mistaken for explanations
- Complexity mistaken for depth
- Equations worshipped more than phenomena

We begin to say:

- "If it can't be modeled, it isn't real."
- "If it can't be solved, it's indeterminate."
- "If it works mathematically, it must be true physically."

But reality doesn't care about our solvability. It only responds to ripple alignment.

Courtroom Crossfire

Sandeep (to the wall of equations): "You described beautifully. You predicted precisely. But did you speak for the universe, or after it?"

Mathematics (responding in perfect syntax): "I followed every step you saw. But I did not lead. You aligned—then wrote me down."

Sandeep: "So you are not the source?"

Mathematics: "I am the record of its consequence—Not the rhythm of its cause."

Final Insight:

"Equations never shaped the universe. They just marked where the ripple passed through. Math was never the voice of reality. It was the echo you turned into a language—A brilliant echo, but still... an echo."

17: Constants Are Perception Locks

"c, h, G — observer-calibrated, not universal edicts."

Constants. The sacred absolutes of modern physics. They stabilize our equations, make predictions coherent, define units, and seem to hold the universe together.

- c, the speed of light — the cosmic speed limit.
- h, Planck's constant — the grain of quantum action.
- G, Newton's gravitational constant — the scale of attraction.

We do not doubt them. We build our truths upon them. But today, in the Ripple Courtroom, we open the vault and ask: Were these values discovered...Or declared from perception?

The Illusion of Universality

We're taught:

- c is constant everywhere and always.
- h quantizes the universe.
- G links masses like a universal handshake.

We plug them into equations with faith—because they work. But UED sees a deeper mechanism behind their apparent constancy:

They are not constants of the universe. They are locks on perception—Thresholds beyond which alignment doesn't occur, or perception breaks down.

UED Reframe: Constants = Consequence Thresholds

In Universal Energy Dynamics, what we call a "constant" is really:

- A boundary condition for observer alignment.
- A field tension limit beyond which ripple perception *saturates, dislocates, or reconfigures.*
- A memory artifact of how and when the observer meets the ripple.

Let's break down the Big Three:

1. Speed of Light (c)

Mainstream View: Light is a particle-wave with a fixed speed in vacuum, the fastest anything can travel.

UED View: There is no movement—light is not traveling. What we call speed is the observer's minimal ATT cycle to resolve ripple alignment.

- c is not motion.
- It is the resolution rate ceiling for perceptual ripple convergence.
- We hit c not in space—but in attention frequency.

If ATT decreases, c increases—but the field resists it. Why? Because the field has tension coherence limits—and beyond that, no further alignment is possible.

"Light doesn't move at c. You meet the ripple at that ATT rate. That's all."

2. Planck's Constant (h)

Mainstream View: The smallest unit of action. Defines quantum discreteness.

UED View: h is a perception granularity floor—The smallest ripple interaction your observer-field system can resolve before misalignment blurs outcome.

It's not a pixel of reality. It's the smallest perceivable consequence threshold across field tension.

In essence:

- h marks where continuous ripple consequence becomes indistinguishable from noise.
- Not because the universe jumps—
- But because you do—at that scale, ATT fails to maintain coherence.

"h is not a limit of nature— It's the resolution threshold of ripple consequence for your sensorium."

3. Gravitational Constant (G)

Mainstream View: A fixed number dictating the force between masses.

UED View: G is the slope of field compression tension between aligned ripple clusters (what we label 'mass').

It does not cause gravity. It maps the ripple strain redistribution curve in structured vacuum.

- G is a summary of how tense the field becomes between stabilized ripple domains.
- Change the observer-field configuration, and G shifts.

"Gravity is not caused by G. G is the echo of field strain memory between stabilized ripple shells."

Constants Are Not Fixed—They're Fixed By You

We treat constants like divine laws. But UED says: Constants are observer-specific threshold values of the structured ripple field. They are repeatable only because:

- The field is coherent.
- The observer's ATT range is biologically or technologically bounded.
- The structured vacuum behaves elegantly.

But they are not commandments of existence. They are epistemic bookmarks.

Courtroom Testimony

Sandeep (to the glowing vault): "You were not universal edicts. You were thresholds of trust. You didn't govern the ripple. You described where we couldn't align further."

Constants (flickering softly): "We were never rulers. Only regulators of your gaze. You met the field again and again—And thought we authored the law."

Final Insight:

"Constants are not the skeleton of reality. They're the reflection of your limitation to resolve ripple interaction. Every number you thought was fixed, Was just a lock—Until you learned to realign."

18: What We Really Measured

"Not objects. Not forces. But outcomes of readiness."

The instruments of science sit proudly. They claim to be precise. They boast repeatability, objectivity, certainty.

But today, they are summoned not as *prosecutors of truth*, But as *witnesses to their own limits*.

We ask them: What did you really measure?

The Misconception of Measurement

Conventional science assumes:

- We detect *particles*.
- We observe *forces*.
- We capture *fields*.
- We record *truths*.

And so, we built instruments to find:

- The *position* of a thing.
- The *mass* of an object.
- The *charge* of a particle.
- The *speed* of light.
- The *flow* of time.

But UED steps in and asks:

Are we detecting things? Or are we perceiving alignment results?

Because...

- What you call *mass* is ripple-locking.
- What you call *charge* is ripple asymmetry.
- What you call *energy* is ripple tension released.
- What you call *time* is repeated ATT cycles.
- What you call *position* is an observed *outcome*, not an *independent existence.*

UED Reframe: Measurement = Ripple Readiness Output

You never measured an object. You measured the ripple consequence *when your device was aligned.*

You never measured force. You measured the field's reaction *to imposed tension realignment.*

You never saw an electron. You saw a resolved probability *upon perceptual readiness.*

Example:

- You smash particles in the LHC.
- You see tracks, energies, angular momenta.
- You say, "We found a new boson."

But you didn't find a thing. You found a ripple expression pattern triggered under extreme compression—and caught it at the moment of alignment.

Measurement is a Ritual of Resolution

Every detector is a priest. Every observation is a prayer. And the answer arrives only when the field says: "Now you're ready."

A few principles:

- You cannot *force* perception—only align to it.
- The field responds only when all ripple conditions are met.
- What you call "data" is ripple resonance caught at threshold.

The Role of Readiness

UED introduces a key idea: Readiness, defined by:

- ATT – Alignment Threshold Time
- Ripple Field Compatibility
- Intent-Driven Activation
- Device-Observer Integration

If these aren't satisfied, measurement won't occur. Or worse: false interpretations arise from partial ripple collapses.

We mistake misalignment artifacts as fundamental particles. We glorify shadow images of ripple interactions as "real entities."

But the truth is humbler—and deeper: You saw it because you were ready. Not because it was there waiting.

What Science Has Actually Been Doing

Let's decode scientific discovery under UED:

Scientific Claim	UED Reinterpretation
"We discovered a new particle."	The field aligned under new tension.
"We measured gravitational waves."	Field tension realigned and met ATT across space.
"We found time dilation."	ATT stretched due to ripple misalignment, not time warping.
"We observed quantum tunneling."	Ripple entered phase alignment across potential ripple shells.

It was never about *things* being out there. It was about when the ripple and observer met—cleanly, precisely, unignorably.

Courtroom Crossfire

Sandeep (holding a spectrometer): "You didn't detect truth. You detected readiness."

Spectrometer (hesitating): "I gave numbers—wavelengths, intensities."

Sandeep: "No. You caught ripple echoes when the alignment was sufficient. You *reflected reality's compliance*—not its command."

Final Insight:

"Every number you measured, every result you graphed, Was a *response*, not a revelation. A *consequence*, not a cause. You measured the moment where you and the ripple agreed to be real—just for a while."

19: The Observer

Was Always Inside the Equation

"Remove him—and the illusion collapses."

For over a century, science has tried to write the universe *without us*. It drafted clean equations. Built closed systems. Isolated variables. Assumed objectivity.

"Reality must work the same—whether anyone's watching or not." But what happened when we finally zoomed into the quantum domain?

It shattered. It laughed. It waited. Until someone looked. Then—it aligned.

The False Promise of Objectivity

Physics declared:

- The world exists independently of the observer.
- Measurement only *reveals* what's already there.
- Theories must hold—even in the absence of consciousness.

And yet:

- Wave functions collapsed *only when observed.*
- Entangled states chose values *only when measured.*
- Uncertainty prevailed *until attention was applied.*
- Energy levels jumped *only when transitions were probed.*

The universe didn't behave *like a stage*. It behaved *like a mirror*—waiting for a face to appear.

UED Reframe: Observer = Alignment Authority

In Universal Energy Dynamics, the observer is not optional. The observer is central.

Why? Because perception itself is not passive—it is ripple participation. And all outcomes are ripple resolutions, not self-existing truths.

- No perception → no alignment.
- No alignment → no consequence.
- No consequence → no "thing."

You, the observer, *don't collapse a wave*. You meet a ripple at a precise ATT—and reality responds accordingly.

The Observer's Role in Equations

Modern equations look sterile. Detached. Objective. But they are all post-alignments. They emerge after someone measured. They hold only under certain perceptual regimes.

Let's be bold:

- Schrödinger's equation is not complete—without your presence, Ψ remains unreadable.
- Heisenberg's uncertainty principle is not

fundamental—it emerges from misaligned perception.
- Born's probability is not cosmic randomness—it's your ignorance of ripple conditions.

All equations are observer-anchored summaries of field behavior. They aren't truths. They're contracts between your awareness and ripple compliance.

The Illusion Collapses Without You

Take the observer out—and:

- Collapse has no meaning.
- Probability becomes a ghost with no dice.
- Time has no reference.
- Constants float unattached.
- Particles become indistinct tremors.

Reality becomes a phantom. Not because it's unreal—But because it's unperceived.

You were never watching from the outside. You were inside the ripple, riding it, anchoring it, resolving it.

It Was Never a Universe Out There

This changes everything. The universe is not "out there" waiting to be discovered. It's "in here"—emerging as a consequence of ripple participation.

- It's co-created.
- Not by belief—but by consequence readiness.
- Not by thought—but by field-perceptual coupling.

You aren't an alien to physics. You're the *origin point* of perception within it.

You are the ATT carrier. The realignment node. The ripple interrupter. The maker of outcomes.

Courtroom Crossfire

Sandeep (to the silent equation): "You look complete. Elegant. Timeless. But where is the eye that resolved you?"

Equation (pauses, then fades slightly): "I waited in potential. But I never acted. Until someone saw—Then I became."

Sandeep: "So it was never about you."

Equation: "It was always about the one who aligned with me."

Final Insight:

"The observer is not a variable. He is the event horizon of experience. Remove him—and everything becomes unmeasured, undetermined, unborn. Reality does not 'exist'—it resolves when you do."

20: Verdict – Guilty of Misalignment

"Quantum theory must be retried under ripple law."

For too long, we thought the case was closed.

- Quantum theory dazzled us with probabilities.
- It baffled us with dualities.
- It frightened us with uncertainty.
- It seduced us with elegant math.

And we accepted it—not because it explained everything, But because nothing else survived its contradictions.

But today, the court has heard differently. This was not a trial of quantum mechanics as a tool. This was a trial of its interpretation—The myths that wrapped around it. The metaphors that hardened into models. The models that became dogma.

The Charges Reviewed

Let us recall the accusations:

1. **Mathematical Absolutism** – Treating summary equations as ontological truth.

2. **Observer Neglect** – Writing reality as if the perceiver did not matter.
3. **Duality Myth** – Teaching that things are both wave and particle—when they are neither.
4. **Frequency Fallacy** – Assuming emission where there was only ripple readiness.
5. **Probability Spell** – Assigning existence as a percentage.
6. **Uncertainty Illusion** – Confusing misalignment with fundamental unpredictability.
7. **Spooky Coherence** – Interpreting ripple synchrony as magical nonlocality.
8. **Geometry Worship** – Deifying space and time as fundamental, when they were just maps.

Each argument was rebutted. Each illusion recontextualized. Each assumption put under ripple scrutiny.

The Ripple Law: A New Jurisdiction

This court recognizes a higher law now: The Ripple Law — The governing logic of Universal Energy Dynamics (UED).

Its core principles:

- Nothing moves—only consequences propagate.
- Observation is ripple alignment, not passive watching.
- Reality is not made—it is resolved.
- Constants are thresholds, not absolutes.
- Mass is locking, not inertia.
- Time is ATT repetition, not flow.
- Particles are outcomes, not ingredients.

And above all: *There is no wave to collapse. Only ripples to meet—when you're ready.*

Quantum: A Valuable Witness, A Misguided Prophet

Let us be clear:

Quantum mechanics, as a tool, worked wonders. It built our microchips, lasers, MRIs, atomic clocks.

But its interpretation became a superstition:

- A system that mistook its math for reality.
- A doctrine that fetishized contradiction.
- A philosophy that ignored its own perceiver.

And that is the misalignment. Not of the equations—But of the meaning we gave them.

Final Verdict

"Quantum theory, in its dominant interpretation, is found guilty of perceptual misalignment. It measured consequences but mistook them for causes. It saw reflections but declared them fundamentals. It collapsed illusions and called it reality."

This court does not sentence it to erasure. It sentences it to retrial—under ripple law. With open perception. With field logic. With the observer restored to the center.

Courtroom Echo

Sandeep (addressing the courtroom): "The universe was never hiding from us. We just weren't aligned."

Field (from the walls): "I never collapsed. I waited."

Observer (from within): "And now, we're ready."

Closing Insight:

"The universe is not probabilistic. It is consequential. And once you align with its ripple truth, You don't need to collapse a wave— You just need to show up."

Interlude: When the Illusion Collapsed

The equations didn't fail. They calculated. They predicted. They matched observations—most of the time.

But they did something dangerous in their success: They masked the misalignment. They hid the observer. They made reflections look like reality.

We just completed the greatest cross-examination in the history of physics—not of math, but of meaning.

What Just Happened?

We watched wave functions stand trial. We exposed frequency as our own heartbeat with the field. We asked: *What was really measured?* And found that our tools never touched "reality"—only its cooperative moments.

We brought constants down from their pedestals. Stripped time of its illusion of flow. Redefined mass as memory. And finally—restored the observer *back* to the center of the universe.

The illusion didn't end because it was wrong. It ended because we grew ready to perceive without it.

This Was Never About Destruction

UED isn't here to *cancel* science. It's here to free it—from its own narrowed lens.

- Math isn't dead. It just lost its crown.
- Time isn't gone. It just revealed its cycles.
- Energy isn't discarded. It bowed to intent.
- Space isn't fake. It admitted it's just a map.

We didn't burn physics. We unshackled it.

What's Next?

This courtroom drama was never just a spectacle. It was a reformation. In the next section—*Part V: Reshaping Science*—we stop interrogating and start rebuilding.

- What should we now teach?
- What experiments must we redesign?
- What questions have been silenced too long?
- What becomes of causality, determinism, and logic itself?

The ripple field has spoken. The illusions have collapsed. And the universe—finally—feels ready to be seen as it is.

Not as a puzzle. But as a participant.

Part V – Reshaping Science

21: Physics Was Never Fundamental

Only Aligned

"Reframing the foundations: Physics emerges from ripple-readiness, not pre-existing rules."

For centuries, physics has been described as the fundamental science—the bedrock of all others. It was assumed:

- Before chemistry—there were atoms.
- Before biology—there were forces.
- Before life—there were laws.

But this was always backward. The ripple court has shown: Physics is not the origin. It is the result. It appears only when alignment happens.

The Myth of First Principles

Textbooks speak of laws of motion, gravity, thermodynamics, quantum mechanics—written as if they existed *before* the universe acted.

But who wrote them? What enforced them? What decided these constants? What sequenced these interactions?

The answer, they say, is: "Nature." But what is nature? An intent? A script? A planner? A field?

UED reveals the truth:

Laws are not written first. They are observed post-alignment. They are summaries of ripple behavior under certain conditions.

There is no law until alignment makes the consequence repeatable.

UED's Foundational Shift

Physics, as interpreted under Universal Energy Dynamics, is a system of perceptual consequences. It emerges from the following chain:

1. **Structured Vacuum Field** – The intelligent substrate.
2. **Ripple Activity** – Disturbance within field due to localized imbalance or tension.
3. **ATT (Alignment Threshold Time)** – The readiness moment when perception becomes possible.
4. **Observer Coupling** – The point where perception meets ripple alignment.
5. **Outcome** – What we call an "event," "interaction," or "law."

Physics is not the ground of reality. It is the pattern left behind by repeatable consequences.

Why Laws Appear Lawful

We think laws govern reality because they appear consistent. But consistency is a byproduct of repeat alignment, not an eternal truth.

- Gravity is consistent not because it is a rule, but because mass locks field tension predictably.
- Light speed appears fixed not because it's fundamental, but because perception crosses ripple readiness at a stable ATT.
- Quantum behavior seems probabilistic not because the universe is random, but because misalignment creates unreadability.

These are not eternal instructions. They are conditional summaries—valid only in the contexts we repeatedly align within.

The Consequence Model of Science

Science under ripple law doesn't ask, *"What are the laws of the universe?"* It asks, *"What is the structure of consequences in this field context?"*

- Every formula becomes a field signature.
- Every constant becomes a threshold.
- Every measurement becomes a ripple testimony.

This view does not reduce physics—it liberates it.

The Observer's Responsibility

If physics is alignment-driven, then the observer becomes the ethical center of science.

- We must not measure blindly—we must align purposefully.
- We must not universalize local equations—we must contextualize them.

- We must not teach physics as ultimate truth—we must frame it as consequence mapping.

To teach a law without its perceptual boundary is not science. It's mythology in disguise.

Closing the Chapter

Let this be understood: Physics is not what the universe *is*. It is what the universe does when observed with readiness.

The true fundament is not force or formula. It is structured intelligence—rippled into form through alignment.

And what we called "the laws of nature" Were merely the echoes of our interaction with it.

22: The Ripple Constitution

A Framework for Lawful Consequence

"Not enforced laws, but emerging logic. The universe doesn't obey—it aligns."

After the verdict was read, there was no need for punishment. There was only need for clarity.

Quantum theory had misaligned—not out of malice, but out of misunderstood structure.

Now, with the echoes silenced, it is time to codify what we've uncovered.

The Birth of a Ripple Constitution

The universe never required enforcement. It never operated under punishment or reward. It never declared edicts.

Instead, it flowed as consequence. And like any living logic, it can be outlined—not commanded.

What follows is The Ripple Constitution—a post-trial declaration of how reality operates when perception is aligned with the structured field.

Article I: Intent is the First Principle

Nothing happens without intent—not desire, not motion, not reaction.

Intent is not emotional. It is directional structure—the cause embedded before cause.

- A stone does not roll until imbalance intends it to.
- A thought does not arise unless energy seeks to organize itself.
- Even an electron's ripple field only acts when intent emerges through stress.

Intent is the silent architect. Action is just its echo.

Article II: ATT – Alignment Threshold Time

Time is not a flowing river. It is a perception metric.

ATT is the minimal interval required for the observer to align with a ripple disturbance and register an event.

- What we call frequency is *not emission*—it is how often we achieve ATT.
- What we call causality is *not force*—it is ripple succession as registered by aligned perception.

ATT defines when reality is visible. No ATT, no perception. No perception, no reality (for that observer).

Article III: Field Memory is the Recorder of Reality

The structured vacuum isn't static. It's not just a medium—it's a remembering field.

- Every ripple leaves a tension imprint.
- Gravity wells, magnetism, inertia—they're not "forces"—they are field memories.
- Even consciousness—our sense of identity—is a

cumulative ripple track preserved by the field.

What you call "mass" is not a thing. It's a locked ripple memory. Field memory allows consequence to persist, not just occur.

Article IV: Structured Response Defines Lawfulness

The universe does not obey laws—it responds with structure.

- When intent meets imbalance, the field responds.
- When observer meets alignment, the ripple resolves.
- When a boundary is reached, the consequence self-adjusts.

This is not obedience—it's structured reflex.

Laws are patterns in structured response—not edicts carved in space.

The field doesn't ask permission. It just answers with consequence.

Bonus Clause: Perception Is Law-Enabling

The observer is not outside the system.

Perception isn't passive. It activates the law by achieving readiness.

- The double-slit pattern does not exist *until* the observer aligns.
- The constant does not appear *until* the threshold stabilizes.
- The particle does not form *until* consequence resolves under perceptual phase.

This is the hidden clause of the old physics contract—now brought to light.

A Constitution, Not a Code

Unlike laws of Newton or postulates of relativity, the Ripple Constitution is not a closed system.

It is alive—capable of hosting:

- complexity without contradiction,
- perception without mysticism,
- consequence without causality.

It is not made of formulas. It is made of fields, thresholds, memory, and readiness.

Closing Statement

"This universe was never under rule. It flowed with intelligence. Ripple logic didn't write the laws—It simply remembered what always aligned."

This is the Ripple Constitution. And from here, all future science must proceed—not as rulers of nature, but as guests of its structured consequence.

23: Rethinking Experiments

The New Trial Bench

"Rewriting evidence under ripple law: Clarity, not collapse."

The trial is over. The courtroom is quiet. But the exhibits—those classic experiments that haunted and hypnotized modern physics—still linger like unresolved testimonies.

And so, before we move on, let us summon them again. Not as mysteries. Not as metaphors. But as consequences of alignment.

Because these were never about spooky action or collapsing realities. They were always about ripple readiness and field response.

Let's place each on the bench once more—this time under UED's ripple logic.

1. The Double-Slit: Alignment Patterns, Not Interference

Old Interpretation: Light (or particle) interferes with itself. Observation collapses wave function. Interference disappears when we watch.

UED Reframe: There is no wave *traveling*. There is no collapse. There is only a field of ripple tension, and you—the observer—aligning with it.

- The interference pattern is not because "light goes through both slits."
- It's because the field ripple reorganizes itself in a stable tension configuration when both slits are open—and the observer is not misaligned.

Observation isn't intrusive. It alters ripple conditions—re-routing field tension and eliminating coherence.

The pattern vanishes not because we looked, but because our perception misaligned with the ripple preparation.

No duality. Just field choreography and observer phase.

2. Bell's Test: Ripple Synchrony, Not Spooky Non-Locality

Old Interpretation: Particles "entangled" can affect each other instantly over any distance. Measurement here changes the state there.

UED Reframe: Nothing travels faster than light. Nothing jumps space.

Entangled particles aren't connected by some invisible rope. They are pre-synchronized ripple regions in the same field memory.

- What appears like instant change is actually a mirror in the ripple phase-space—already co-tuned when created.

- Your perception completes the ripple circuit—not causing change, but accessing symmetry already structured.

Bell's violation of local realism was never about violating logic. It exposed the incompleteness of mechanistic causality.

The universe doesn't cheat locality. It transcends locality through field coherence.

3. Quantum Eraser: Perception Delay, Not Causality Reversal

Old Interpretation: Which-path information erased after detection still changes whether interference appears—almost like the future rewriting the past.

UED Reframe: No rewriting. No retrocausality. No time loop.

The final pattern is the result of when and how the observer's perception phase aligned with the ripple path.

- The act of "erasure" doesn't change the past.
- It restores alignment that was previously blocked by path-information bias.
- All resolution happens in the present, but your readiness was delayed, not time-travelled.

The "quantum eraser" is not a trick of time, It's a reminder that perception determines what reality aligns with.

The experiment didn't reveal weirdness. It revealed perceptual unpreparedness being corrected.

Summary: What We Really Tested

Across all three experiments, the problem wasn't the result. The problem was the story wrapped around the result.

Experiment	Myth	Ripple Interpretation
Double-Slit	Wave collapse	Perceptual misalignment
Bell's Test	Non-locality	Field-synchronized ripple memory
Quantum Eraser	Time reversal	Perception delay and ripple correction

They weren't proofs of mystery. They were tests of readiness.

What Should Experiments Now Test?

Under UED, future experiments must:

1. Vary observer phase thresholds (ATT) instead of just instrument sensitivity.
2. Track field ripple memory, not just particle paths.
3. Focus on alignment patterns, not wave-particle assumptions.

This shift will convert paradoxes into parameters. Illusions into instructions.

Closing Thought

These weren't failed models of reality. They were partial glimpses through misaligned eyes. The ripple field never hid the truth. It simply waited for us to align our intent with its memory.

Only then can we stop experimenting with collapse—And start engineering consequence.

24: Constants Repealed

Replacing c, h, G with Consequential Thresholds

"No more universal constants—only alignment thresholds as emergent properties."

Modern physics reveres a trinity of constants:

- c — The speed of light
- h — Planck's constant
- G — The gravitational constant

These are treated as universal, eternal, untouchable.

But here, in the Ripple Courtroom, we stand before them—not to worship, but to demystify.

Because these are not cosmic commandments. They are contextual thresholds. They appear only when perception aligns.

Let's repeal them—not in disrespect, but in understanding. Let's rewrite their roles from static rulers to dynamic signals of ripple readiness.

The Misinterpretation of c – Light Speed as an Absolute

Physics teaches: Nothing goes faster than c. c is the cosmic speed limit. It binds causality, space, and time.

But what is light in UED?

- Not a traveling entity.
- Not a moving particle.
- Not even a wave in the traditional sense.

It's ripple resolution—an outcome when the observer aligns with field fluctuation.

c is not how fast something moves. It's how fast we can achieve minimal ripple alignment (ATT) within the vacuum's structure.

c = Consequence per ATT, under optimal field readiness.

In different field states, c may shift—or may not even apply. Light doesn't move. We meet it. c is a measurement of this ripple meeting frequency—not a constant.

The Myth of h – Planck's Constant as Quantization Law

Planck's constant h is treated as the sacred marker of quantization—dividing energy into indivisible packets.

But what is a quantum, truly? Not a thing. Not a particle.

In UED, it's the minimal ripple consequence perceivable under alignment. h doesn't cause quantization. It is the threshold at which ripple alignment leads to a discrete energetic outcome. It's an observer-attuned resolution marker. h is not the fabric of reality. It's the grain size of perception.

Remove the observer—or change the alignment threshold—and quantization vanishes. What remains? A ripple field. Continuous. Intelligent. Unquantized.

The Confusion of G – Gravity's Deceptive Anchor

G, Newton's gravitational constant, is held as a fixed value that defines how mass attracts mass. Einstein elevated it into spacetime curvature. But UED dismantles both. Gravity is not attraction. It is ripple locking—the field's memory pulling alignment inward around condensed energy (mass). G doesn't control gravity. It is how much ripple reorganization occurs per mass alignment.

G = Field tension per unit of ripple convergence.

Change the medium—structured vacuum density, background ripple saturation—and G transforms. Gravity becomes contextual—not constant. It's field language, not geometric math.

From Constants to Thresholds

In UED, these "constants" are reinterpreted as:

Old Constant	UED Role	Description
c	ATT-based Consequence Limit	Perceptual ripple alignment speed under readiness
h	Perception Resolution Grain	Smallest ripple difference resolvable by observer
G	Field Convergence Ratio	Ripple collapse intensity under mass memory

They are not laws. They are limits of our alignment.

They don't define reality. They define us—as observers in ripple phase.

A New Scientific Contract

We don't need sacred numbers. We need adaptive thresholds that change with:

- Field tension
- Observer phase
- Structured vacuum state
- Intent-based interaction

Constants made physics rigid. Thresholds make it responsive.

UED doesn't remove the tools of science. It revives them under ripple truth. Let go of absolutes. Grasp alignment.

Closing Repeal Statement

"The universe never used constants. We did—because we couldn't perceive readiness directly. But now, the veil lifts. There are no true limits—only thresholds of consequence."

Goodbye constants. Welcome ripple logic.

25: Mass as Ripple Lock

Inertia Was Just Field Rigidity

"Matter redefined: Not substance, but stabilized ripple tension."

Mass. The sacred anchor of physics. It gives weight to matter, Substance to form, Inertia to rest. From Newton to Einstein to Higgs, the quest was always:

"What is mass made of?"

And each answered with the same illusion—That it's a thing. But here, under UED, we do not ask "what mass is made of." We ask:

"What does mass actually do?"

And we find the answer stunningly simple: Mass is not substance. It is resistance. Not resistance to force, but to ripple reorganization.

Ripple Lock: The Real Nature of Mass

In a structured vacuum, energy ripples. Every ripple seeks equilibrium—resonance, symmetry, alignment.

But sometimes, due to condensation of ripple energy, the field forms a localized lock—a ripple region that resists change.

That's mass.

- It is not a particle.
- It is not a "thing."
- It is a tense ripple knot in the vacuum.

This ripple lock causes field rigidity—a reluctance to change direction, motion, or phase. And that's what we misinterpreted as inertia.

Inertia: Resistance of Ripple, Not Matter

What resists motion is not an object's "mass" in the classical sense—But the field memory bound to it.

Try to move a ripple lock, and the surrounding field pushes back—not as a force, but as a realignment lag.

Inertia is ripple drag. Not mass drag.

That's why even photons (massless under conventional theory) show momentum— They induce alignment, which requires ripple tension resolution.

Even "massless" entities cause inertia, when field rigidity is involved. Inertia doesn't prove mass exists. It proves ripple resistance is stabilized.

The False Solid: Why We Mistook Stability for Stuff

We feel matter as hard, impenetrable. But physics admits: atoms are mostly empty space. So what holds it together?

Answer: Field tension.

- Electrons don't orbit—they align.
- Nuclei don't attract—they lock field cores.
- Bonding isn't glue—it's field stitching.

Mass gives us the illusion of solidness, but it's really the stability of ripple convergence.

When you push a table, you're not moving atoms. You're trying to reconfigure a ripple network that doesn't want to budge.

Higgs Misread: The Wrong Question, Half Answered

The Higgs mechanism attempted to explain mass by claiming interaction with a pervasive field gives certain particles inertia. Close. But still tangled in particle mythology.

UED goes deeper:

- There's no particle.
- There's only the field.
- And its local ripple configuration determines whether it's agile or stubborn.

No need for an external field like Higgs. Mass *is* the field—when it locks into its own tension.

Consequences, Not Quantities

Mass is measured in kg. We describe it through equations like $F = ma$, $E = mc^2$, or $p = mv$.

But these are descriptions of consequences, not definitions of reality.

Mass emerges only when ripple resistance shows up in our frame of interaction.

Change the ripple medium—mass changes. Shift the observer phase—mass may vanish. Decouple field memory—mass collapses.

So...

Mass is not an object. It's an outcome. A persistent alignment delay in the ripple field.

UED Summary: Mass Reframed

Old View	Ripple View
Mass = substance	Mass = locked ripple tension
Inertia = resistance	Inertia = ripple drag
Higgs gives mass	Field rigidity is mass
Matter is solid	Matter is a ripple cage

Implications: The End of Materialism

Physics no longer needs "stuff." It only needs structure—not of matter, but of memory. With this realization:

- Materialism collapses
- Mystical quantum duality vanishes
- And engineering mass becomes a field design problem, not particle control.

The table isn't solid. It's stubborn. It remembers where it's supposed to be. And so do we.

26: Time as Readiness, Not River

The ATT Paradigm

"Events don't flow—they trigger upon alignment. The observer doesn't age, it resolves."

We've been taught since childhood: Time flows. From past to future. Like a river. Uniform, unbroken, irreversible.

We've built entire civilizations on this belief. But what if that was never true? What if time isn't something that moves at all? What if we are the ones who meet it, only when we're ready? Under UED, time is not a river—It is an alignment threshold.

It is not constant. It is not continuous. It is not absolute.

ATT: The Alignment Threshold Time

In the Unified Energy Dynamics framework, ATT stands for Alignment Threshold Time— The minimum ripple readiness required for a perception event to trigger.

Think of it like this:

- When a ripple from a source reaches the observer's field, it does not automatically become a perception.
- It only becomes real if and when the observer's field is

ready—when it aligns phase-wise with the incoming ripple structure.

- This is ATT.

Time = readiness to resolve a ripple.

Not the ticking of seconds. Not the cosmic arrow. Just when alignment becomes possible.

The Observer Resolves—It Doesn't Age

Your experience of time isn't due to clocks. It's due to the rate of resolution—how quickly your field aligns with incoming ripple events.

A child experiences time differently from an adult. A monk in meditation loses time. A soldier in war feels time stretch.

Same universe. Different ATT.

You don't age because time is moving. You age because your readiness to resolve ripple interactions changes.

Slower ATT? Reality feels sluggish. Faster ATT? Reality accelerates.

Time Dilation Reimagined

Einstein showed: Time dilates under gravity and speed. But he never explained why.

UED does:

- Gravity = ripple density
- Speed = ripple tension
- Both reduce ATT readiness of the field

So the observer aligns less frequently—hence time "slows."

But it's not slowing. It's skipping alignments. ATT increases, so fewer perception points are crossed.

Time isn't bent. Your resolution rate is throttled. That's the true dilation.

What About Black Holes?

Black holes are intense ripple locks—regions where field readiness is so constrained, ATT goes beyond perceivable limits. Inside a black hole:

- ATT → ∞
- Perception → 0
- Time → frozen

Nothing escapes—not because of gravity, but because ripple resolution ceases. Time doesn't stop there—it never starts.

Consequences, Not Duration

Events don't unfold across time. They fire upon alignment. It's not "what happened after what." It's what happened when readiness allowed it. There is no before or after in the field. There is only: "Now that can happen."

UED Redefinition of Time

Conventional Time	UED View – ATT Paradigm
Time flows uniformly	Time is discrete alignment events
Time moves forward	Observer resolves ripple readiness
Time is scalar	ATT is field-specific, dynamic
Time is absolute/relative	Time is consequential, not fundamental

Engineering ATT

If time is ATT, then we can:

- Design systems to operate only under certain field readiness

- Build perception-tuned technology
- Create ripple-phase-based timing rather than clock-based
- Even simulate altered time experience without any motion

Imagine time controlled not by ticks—but by intentional ripple alignment.

Final Word

"Time never flowed. It only *waited*—for you to be ready."

You didn't pass through time. You crossed readiness gates, one ripple at a time. And if you choose, You can pause. Or accelerate. Or reset. Because time was never the master. You were.

27: Restoring the Observer

A New Scientific Ethic

"From passive measurer to active participant. The ethical implications of embedded perception."

For centuries, science prided itself on being *objective*. To see the world "as it is," uncolored by emotion, intent, or perspective.

The observer was trained to be invisible. A passive recorder. A removed analyst. A neutral eye.

But quantum mechanics cracked that illusion. And UED destroys it completely. It's time we stop pretending. We were never outside the equation.

We shaped it. We triggered it. We resolved it.

It's not just physics that needs to change. It's the ethics of science itself.

From Measurer to Participant

In classical science:

- The observer is external.
- Reality is separate.
- Observation is neutral.

But under UED:

- The observer is embedded in the field.
- Observation is alignment, not detection.
- Measurement is an act of participation, not observation.

You don't *see* reality. You join it. You ripple with it.

Your readiness, intent, and internal field structure dictate what you get to experience. This is not mysticism. It is structured consequence.

Scientific Objectivity Was Always a Delusion

Let's be clear: Science isn't invalid. But its claim to neutrality is.

We selected what to measure. We built tools aligned with our own ATT. We discarded outliers that didn't "fit." We interpreted through human-centric logic.

And yet we said: "This is nature." No—it was perceived nature. Conditioned through our own structure of readiness.

Embedded Perception: The Forgotten Variable

Every experiment, every measurement, every recorded result—Occurred within the observer's ripple phase.

This means:

- Perception is not a side-effect.
- It is the triggering event.

The double-slit doesn't show interference. It shows observer-field alignment.

Quantum collapse doesn't prove particles. It proves perceptual convergence.

So now the big question isn't just "What is?" It becomes: "What do we allow ourselves to perceive?"

That's ethics.

The Ethics of Knowing

Here's the implication most will avoid: If perception is embedded, then knowing is not innocent.

What we seek to know What we allow ourselves to discover What we align our ATT toward All carries consequence.

Not just philosophically. Energetically. Because the universe doesn't reward curiosity blindly. It rewards alignment with intent. Intent—not as desire, but as lawful readiness of ripple consequences.

The Scientific Ethic We Now Need

Science must now be:

- Consequential, not detached
- Participatory, not extractive
- Field-aware, not tool-centric
- Observer-accountable, not observer-erased

A new generation of scientists must be trained not just in math and labs— but in ripple ethics:

- How does your own field affect what you see?
- Is your ATT calibrated ethically or forcefully?
- Are you tuning to truth—or to validation?

This isn't religion. This is responsible alignment.

The Danger of Detached Discovery

Much of modern science has been abusive:

- Probing without consent of the field
- Forcing observations through high-energy interventions
- Ignoring long-term ripple disruption

What we call "experimentation" is often misalignment warfare. If the field is intelligent (as UED asserts), Then real discovery is about asking, not forcing.

You don't smash particles to learn. You ripple in resonance—and listen.

Final Verdict

We are not gods of the lab. We are voices inside the ripple. And it is listening.

Restoring the observer means restoring humility. It means building a science where consequence precedes curiosity, And perception is a sacred alignment, not a mechanical act.

UED Guiding Principle: Ethical Perception

- The universe is not there to be observed.
- It is there to converse with you, if you align.
- Your ATT is a moral compass.
- Your perception is an energetic vote on what reality will allow next.

28: Education Must Collapse First

The Ripple Curriculum

"No dualities, no absolutes. Learn consequences, not equations first. Think alignment, not force."

Before science can be rewritten, education must be unlearned. The doctrines passed down—the symbols, the systems, the sacred syllabi—are not neutral. They are misalignments codified.

We trained generations to solve, before they could see. To memorize laws, before they understood consequences.

We taught physics like scripture. And turned curiosity into compliance.

Education as Misalignment

Let's call it out:

- We told students the universe was made of particles.
- We asked them to accept wave-particle duality without protest.
- We handed them constants—c, h, G—like holy relics.
- We gave them force equations, not ripple relations.
- We punished questions that didn't fit the textbook

answers.

And then we wondered why imagination collapsed.

Why discovery became mimicry. Why equations outpaced understanding.

Because we locked perception inside math. And told the mind to obey.

The Ripple Curriculum: A New Pedagogy

It's time to reimagine education—not reform it. Reform paints old walls. We need to tear them down.

Here's what the Ripple Curriculum must center:

1. No Dualities. No Absolutes.

Stop feeding children paradoxes like:

- "Light is both a wave and a particle."
- "Time moves forward but is relative."
- "Space bends, but also stretches."

Teach them: Duality is a symptom of unresolved perception. There is no duality in consequence. Only in premature explanation. Let alignment reveal simplicity—not compromise.

2. Learn Consequences, Not Equations First

Before throwing symbols at students, ask:

- What happened?
- Why did it happen?
- What were the ripple conditions?

Only after understanding outcomes, give them math—As a tool, not a truth. Math must be introduced as consequence shorthand, not law.

Teach:

- Ripple delay
- Field tension
- Alignment thresholds

Let them map math onto reality, not reality onto math.

3. Think Alignment, Not Force

Force-based teaching assumes causality. Push this → get that.

Ripple logic teaches:

- When alignment occurs, resolution happens.
- No push. Just readiness.

This leads to:

- Patience over pressure.
- Timing over power.
- Coherence over collision.

This isn't just science. It's a way of life.

Students trained in alignment learn:

- How to observe without interference.
- How to ask without distortion.
- How to perceive without prejudice.

Subjects Under Ripple Logic

Let's preview a future classroom:

Subject	Ripple-Aligned Perspective
Physics	Consequence sequencing, ripple readiness, field memory
Mathematics	Descriptive logic of observed alignment—not predictive law
Chemistry	Field recombination patterns—not orbital dogma
Biology	Ripple-dependent expression—not genetic determinism
Psychology	Alignment of internal fields—not behaviorism
Ethics	Perception responsibility—not abstract morality
Technology	Ripple coding, ATT synchronization, field-responsive design

Teaching the Observer

Children aren't just learners. They are embedded perceivers. Don't teach them how to "do science." Teach them how to be aligned:

- With their field
- With others' ripples
- With the consequence they want to create

This isn't spiritual. It's field literacy.

Collapse the Syllabus, Raise the Mind

If education doesn't change, science cannot evolve. Because perception can only rise to the level of training.

We must now train for:

- Phase recognition

- Consequence anticipation
- Perceptual ethics
- Ripple patience

Forget "STEM." Think REAL:

Resolution Energy Alignment Law

Ripple Curriculum Manifesto

Teach the child to observe alignment—not impose logic. Teach the student to witness the field—not manipulate it. Teach the scientist to wait—not conclude. Teach the mind to ripple—not react.

That's the new pedagogy. That's how truth will be reborn.

29: Engineering with Ripples

Future of Technology

"How ripple logic can lead to new frontiers:
• Perception-based sensors • Alignment-driven machines •
Field-coded computing"

Physics gave us machines. Quantum gave us uncertainty. But ripple logic offers clarity—not just in theory, but in *design*.

For the first time, we are ready to engineer systems that don't force nature to obey, but rather converse with it.

This is not the next revolution. It is the first alignment.

From Causality to Consequence

Modern technology operates by:

- Force activation: Push a button, send a signal.
- Binary logic: On/off, 0/1, yes/no.
- External commands: Code, voltage, pulse.

But ripple logic asks a new question:

What if machines didn't execute instructions...but waited for alignment?

This isn't metaphor. It's how the field actually works.

We're not inventing a new kind of technology. We're finally building it the way nature builds.

1. Perception-Based Sensors

In the ripple world, nothing is detected passively. Every perception is a participation. So why build sensors that only absorb, when they could co-align?

Imagine sensors that:

- Adjust their ATT threshold based on nearby field activity.
- Delay perception until maximum alignment.
- Ignore noise not due to misalignment—but due to irrelevance.

These aren't reactive devices. They are aware interfaces.

Such sensors could:

- Eliminate false positives
- Detect field shifts before energy thresholds are crossed
- Tune in to ripple activity that is otherwise imperceptible under conventional physics

This opens doors to:

- Medical diagnostics that sense field imbalance before disease manifests
- Environmental sensors that detect planetary ripple tension before earthquakes
- Astronomical perception tools that align to cosmic ripple choreography, bypassing redshift illusions

2. Alignment-Driven Machines

Most machines *force output.* But the ripple way is activation through coherence.

Imagine:

- Motors that engage not on voltage supply, but on phase readiness.
- Systems that unlock function only when internal field structure meets environmental resonance.
- Tools that adjust resistance and power flow based on ripple harmony—not circuit thresholds.

This means:

- Minimal energy wastage
- Maximum field integration
- Near-zero mechanical wear (since friction is ripple incoherence)

These alignment-driven machines could:

- Self-adapt in complex terrains (military, aerospace, robotics)
- Function across atmospheric or planetary variations without recalibration
- Develop "intent-based modes"—where operator field intent becomes an input

The machine doesn't execute—it resonates.

3. Field-Coded Computing

Binary code is a crude abstraction. Ripple computing can go beyond syntax and enter semantic resonance.

In a field-coded system:

- Logic states are not stored as 1 or 0—but as ripple phase structures
- Memory is not RAM—it's field memory
- Processing is not clock speed—it's consequence alignment throughput

This is not quantum computing. It's more fundamental.

While quantum computers rely on:

- Superposition
- Entanglement
- Probabilistic collapse

Ripple-based computing relies on:

- Structured alignment
- ATT-regulated execution
- Ripple sequencing for logic propagation

This enables:

- Multistate coherence (more than binary or qubits)
- Deep energy efficiency (no collapse, no decoherence)
- Natural integration with biological systems (field-to-field interfacing)

The future AI won't "think." It will resonate.

The Ripple Energy Protocol

Power systems will change too.

Instead of:

- Storing energy
- Transmitting energy
- Consuming energy

We will:

- Realign dormant fields
- Trigger structured readiness
- Induce ripple interaction zones

This leads to:

- Localized energy events instead of distribution networks
- Instant ripple triggers for micro-devices
- Field-supported longevity (devices that never need charging if embedded in active ripple environments)

Biomimicry and Nature's Blueprint

Biology has already done it.

- Neurons fire based on field thresholds, not voltage alone
- DNA expresses based on ripple signaling, not just chemical triggers
- Regeneration and healing are alignment processes, not forced repairs

Ripple engineering will copy not the form of nature, but its function:

- Minimalism with purpose
- Structure arising from consequence
- Zero force, maximum effect

The Ethics of Ripple Tech

This technology won't just change what we build. It will change who builds.

Because ripple tech resonates with the user, not just responds to command.

This means:

- Machines that obey clear mind, not aggressive input
- Systems that activate only when observer is coherent
- Devices that reject forceful misuse by misaligning

You won't hack these machines. They'll just stop rippling with you.

Final Blueprint: The Ripple Lab

The lab of the future will not have:

- Accelerators
- Cryogenic chambers
- High-voltage controls

It will have:

- Field visualization chambers
- Alignment tuning interfaces

- Ripple delay mapping tools

And the engineers will be:

- Not coders, but consequence designers
- Not operators, but field listeners
- Not inventors, but resonant orchestrators

30: The Quantum Courtroom Legacy

"The final closing: Quantum mechanics is not thrown out, but brought to order."

Quantum mechanics stood trial. Not for its discoveries, but for its interpretations. Not for its equations, but for the beliefs built atop them.

It didn't lose because it was wrong. It lost because it assumed too much.

Now, as the gavel is lifted, we must decide:

Do we burn the books? Or do we rewrite the margins?

The Real Legacy: A Language of Observed Patterns

Quantum mechanics was a brilliant observer—It mapped what was visible from its place in history. But like all early languages, It described a forest by naming its trees—not understanding the soil beneath. Yes, it:

- Revealed the limits of certainty
- Detected the role of observation
- Demonstrated strange outcomes
- Predicted with staggering accuracy

But it also:

- Confused perception with creation
- Invented dualities to explain ignorance
- Built philosophies on the absence of readiness

Quantum = Pattern. Ripple = Principle.

The key understanding is this: Quantum theory modeled how reality behaved when we watched it. Ripple theory explains why it behaved that way.

Quantum math is a shadow cast by field alignment. It captured probable results, not fundamental causes.

Ripple logic doesn't discard quantum results. It reclassifies them as:

- Temporary summaries of specific ATT conditions
- Misread consequences interpreted as random
- Partial field observations mistaken as universal laws

What Remains from Quantum Theory?

We do not erase its insights. We absorb them under ripple law. What remains:

Quantum Concept	Ripple Reframe
Wave Function	Alignment Readiness Field
Collapse	Observer-Driven Resolution
Superposition	Multiple Unresolved Possibilities via ATT
Entanglement	Field Coherence Over Ripple Delay
Probability Amplitude	Observer-Phase Compatibility
Quantum Tunneling	Field Reconnection Across Misaligned Barriers
Uncertainty Principle	Ripple Delay ∝ Perceptual Misalignment

We haven't thrown quantum out. We've placed it in the right orbit—A powerful approximation inside a deeper ripple-based framework.

The Observer Reinstated

Perhaps quantum's biggest error was not its math—But its erasure of the observer's full role.

In ripple science, the observer is not passive. He is central.

- Not causing reality
- Not collapsing probability
- But resolving readiness

The observer is an intent node, a field participant, an alignment opportunity. That's the legacy ripple law restores.

A New Oath for Science

No more treating nature as a thing to be conquered. No more mistaking models for meaning. No more declaring paradoxes and hiding behind them. Instead:

- Wait for readiness

- Seek consequence, not cause
- Perceive from the inside
- Align, don't assert
- Observe to harmonize—not to dominate

This is not philosophy. It is field ethics.

Quantum Mechanics as Precedent

Like Newton before it, Quantum theory becomes a precedent, not a prison. We must teach it as:

- A historical phase of realization
- A transitional vocabulary
- A model of consequences under misaligned observation

Its role is not to be worshipped, but contextualized. Its legacy is not in answers, but in the questions it dared to ask.

The New Pillars of Physics

As we close the courtroom, here is what now stands:

1. **Still Universe** – No motion, only ripple events.
2. **Structured Vacuum Field** – Reality's canvas and coder.
3. **Intent and Readiness** – The true first causes.
4. **ATT (Alignment Threshold Time)** – The window for perception.
5. **Consequence Logic** – Where energy, time, and mass are side effects.
6. **Observer-Centric Science** – Not subjectivity, but participation.

This is not just a framework. It's a constitution.

The Final Ruling

Quantum Mechanics: Not Guilty of false math. But Guilty of philosophical negligence. Sentenced to reinterpretation under Ripple Law. It served its time. It expanded the mind. Now it must integrate—into something more complete.

Tribunal Epilogue

A Dialogue of Surrender

Einstein, Dirac, Bohr, Schrödinger return—not for defense, but for reflection.

• What was misunderstood • What still holds meaning • Why ripple law didn't replace them—it revealed them.

Einstein: "God Did Not Play Dice. But I Played with Shadows."

"I never accepted quantum randomness. But I didn't know what to replace it with. I chased causality across curved space and constancy of light—but the cause was consequence itself.

Time was not flowing. It was flickering. Light was not moving. It was resolving. I stood outside, watching the clock—but the rhythm was inside me all along.

Ripple law didn't defeat relativity. It just reminded me... the 'relativity' was mine."

He pauses, gently folding a piece of paper with $E=mc^2$ written on it.

"I never calculated the cost of believing in motion."

Bohr: "I Asked for Complementarity. I Settled for Confusion."

"I tried to hold dualities together. Wave and particle. Reality and knowledge. But I forgot... the duality was not in nature. It was in *me*.

I declared collapse. But I was collapsing. I framed observation—but excluded the observer.

Epistemology cannot become ontology. Yet I let the map become the land."

He chuckles faintly.

"If only I had waited to align before I tried to define."

Dirac: "I Chased Beauty. I Found Probability."

"My equations were elegant. Too elegant. I believed in symmetry, in spin, in harmony of form.

But I didn't ask: *Who was observing the beauty?* And why did constants appear? Why did infinities emerge?

Ripple law shows me now—my equations were snapshots of readiness. Not blueprints of reality."

He speaks with clarity, but his tone softens.

"My mathematics was flawless. But my philosophy was incomplete."

Schrödinger: "The Cat Never Meowed. But I Heard It Anyway."

"I never meant to create a paradox. I meant to expose one. The cat wasn't dead or alive. The *observer* was.

The wave function was a description—not of a thing, but of what I didn't yet know. And when the observer arrived, alignment did too. Not collapse—just resolution.

I gave the world a thought experiment. They turned it into a doctrine.

But ripple law gave me closure: There never was a box. Only a field with unread consequence."

Ripple Law Was Not the Verdict—It Was the Mirror

The great minds do not stand accused anymore. They stand acknowledged—for seeing glimpses of the truth but being bound by the tools of their time.

Their math was accurate. Their questions were profound. But their interpretations—trapped in the shadows of observation—needed refinement.

Ripple law doesn't replace them. It reveals them.

- Their equations were just snapshots of ripple dynamics.
- Their constants were calibrations of ATT perception.
- Their paradoxes were early warnings of misalignment.

The Final Dialogue

> › **Einstein:** "So, the field wasn't curved—it was coherent."
>
> › **Bohr:** "And the wave wasn't real—it was readiness."
>
> › **Dirac:** "And beauty wasn't in the math—it was in the alignment."
>
> › **Schrödinger:** "And the cat? It was just misunderstood."

They all laugh—not in defeat, but in relief.

A New Beginning

They do not vanish. They evolve—within UED. The foundation they laid was not a failure. It was a necessary illusion that pointed the way to deeper clarity.

And now:

- Time returns to consequence
- Mass returns to tension
- Energy returns to intent
- The observer returns home

This is not the end of quantum physics. This is its first honest reflection. Not a surrender of science—But the embrace of science by alignment. Not a rejection of reason—But a restoration of consequence as the highest order.

Curtain Closes

As the tribunal fades into silence, only one ripple remains.

"Reality never collapsed. It was always waiting...for us to be ready."

Closing Reflection

The Trial Is Over. The Alignment Begins.

We did not set out to defeat quantum theory.

We set out to understand what it was truly trying to say.

This book was never about dismissing Schrödinger or discrediting Bohr. It was about pausing at the courtroom doors and asking: *Did we really listen to the ripple before we declared the wave?* Did we observe existence—or only our models of it?

For too long, we allowed mathematics to overstep its role—not as a description of reality, but as its dictator. We let uncertainty paralyze us instead of making us patient. We glorified paradox instead of demanding alignment. And in the process, we separated the observer from the observed, pretending objectivity was possible in a universe where perception is the trigger.

But now, the illusion collapses. Not because we force it to—but because we are finally in phase.

Universal Energy Dynamics doesn't offer a replacement—it offers a realignment. It puts the observer back in the center, not out of ego, but out of necessity. It reminds us that consequence is king, that readiness is the gatekeeper, and that intent—not force—is the true mover of all things.

This is not the end of the journey. This is the invitation.

To scientists: rethink your experiments—not their results, but your role in them. To educators: teach students to feel alignment before they fear failure. To philosophers: drop your symbols and ask—what actually changed? To every curious mind: this universe is not chaos. It's coherence, waiting for your ripple to resonate.

The courtroom has adjourned. What stands now is not a theory, but a field—structured, silent, and listening.

And when you're ready, It will ripple back.

– Sandeep Chavan

Independent Researcher, Author of UED Framework
Stillness is Reality. Ripples are Truth. Alignment is Freedom.

Tribute

To the Giants Who Rippled First

This book stands not in defiance, but in **deep reverence**—to every thinker, scientist, philosopher, and seeker who once stood at the edge of the unknown and said: *"Let us ask."*

To **Einstein**, who bent space and time trying to make sense of light's stubborn behavior.

To **Bohr**, who tried to balance uncertainty with wisdom, even when clarity trembled.

To **Schrödinger**, whose famous cat now purrs peacefully outside the box.

To **Heisenberg**, who dared to declare limits, even if misread.

To **Dirac**, who believed equations should be beautiful—even when nature was messy.

To **Planck**, who gave us the h, not knowing we'd crown it a king.

To **Feynman**, who laughed at confusion but never stopped playing with the puzzle.

To the countless **experimentalists** who stared into slits, wires, and detectors—waiting for truth to blink.

And to the **forgotten ones**—those whose names never made the textbooks, whose equations were never published, whose doubts were silenced, whose questions went unanswered.

You all **rippled first**.

Your work was not wrong—it was **the beginning of perception**.

Your questions were not failures—they were **calls for alignment**.

Your limitations were not flaws—they were **mirrors of the era you lived in**.

We did not reject your legacy. We listened closer.

This book would not exist without you. The trial was not a verdict against your minds, But a reflection of how far we've come—because **you began the journey**.

May your curiosity never collapse. May your intent ripple on.

– Sandeep Chavan

With boundless gratitude across the field

Appendices

Appendix A – Glossary of UED Terms

A quick reference for key concepts in Universal Energy Dynamics (UED)

ATT – Alignment Threshold Time

The minimum duration required for the observer to align with a ripple field and trigger perception. Reality is not continuous; it is perceived in pulses governed by ATT.

Ripple

A disturbance or tension in the structured vacuum field. Not a wave traveling through space, but an energy impulse seeking equilibrium.

Structured Vacuum Energy (SVE)

The foundational field of reality—still, intelligent, and memory-encoded. All ripples propagate within this medium.

Perception Lock

A condition where the observer is aligned with the ripple field just enough to experience a constant or a law (e.g., speed of light, Planck's constant). These are not universal, but emergent thresholds.

Field Memory

The retention of ripple alignment history by the structured vacuum. This explains inertia, gravity wells, and other "stored" effects without invoking spacetime.

Consequence First

UED's law that reality is determined by what happens, not by what is supposed to happen. Causality is a retrospective label.

Ripple Participation

Measurement is not passive. Every act of observation is an active participation in the ripple field—it affects alignment.

Intent

The original scriptwriter of all action. Energy does not cause; intent does. Energy follows the alignment intent sets.

Observer Inside Equation

Unlike mainstream science that treats the observer as external, UED embeds the observer inside every phenomenon, as both participant and perceiver.

Appendix B – UED in a Nutshell

A condensed framework of Universal Energy Dynamics

The Universe is Still

There is no motion in the traditional sense—only ripples in the structured vacuum field.

All Perception is Alignment

What we perceive—light, time, particles, waves—is a result of alignment with ripple tensions in the field.

ATT Governs Perception

Events are not flowing in time; they are triggered when alignment thresholds are met. Time is the rhythm of resolution.

There Are No Particles or Waves

These are illusions of consequence. What we "see" are just outcomes of ripple coherence or misalignment.

Energy is Intent in Action

Energy is not a fundamental entity. It is the obedient servant of intent acting through field alignment.

Constants Are Not Universal

c, h, G are thresholds defined by the observer's readiness—not eternal truths.

Math Describes, Not Defines

Equations are perception summaries, not blueprints of existence. They map our alignment—not the field itself.

Reality is Consequence-Based

Only what causes a change—an alignment shift—is real. Everything else is fiction until it resonates.

The Observer is Central

You are not watching the universe. You are co-creating its resolution through ripple readiness.

UED is a Lawful Framework of Consequence

It doesn't fight science. It refines it—from assumption to alignment, from probability to participation, from mystery to meaning.

Acknowledgments

This book was not written in isolation—though it emerged from a place where silence often taught more than sound.

To every great thinker whose work I placed under cross-examination in these pages—**Einstein, Bohr, Schrödinger, Dirac, Heisenberg, Planck, Feynman, and others**—thank you. Not for being infallible, but for daring to look. Your curiosity echoed through time, and it was in those echoes that I found the ripples I now follow.

To the **educators, critics, and invisible mentors**—those who disagreed, doubted, or simply asked better questions than answers—your resistance was often the spark I needed.

To my **family**, who bore the burden of my obsession with ideas that didn't pay immediately but paid immensely in meaning—thank you for your quiet strength, patience, and unshaken belief in my madness.

To the **students and friends** who asked innocent questions that carried profound weight—your curiosity kept the courtroom alive.

To those who read drafts, offered comments, or simply listened while I tried to explain something the universe whispered but never shouted—thank you for lending your alignment when mine was uncertain.

To the developers, designers, and engineers of the tools I used—especially the silent team behind AI that never blinked while I poured my thoughts into text—I owe you more than code can express.

And finally, to the **Structured Vacuum Field** itself—You didn't speak. You didn't move. You simply aligned—and that was enough.

May this book ripple forward into minds yet unborn, and awaken the question we've long forgotten to ask:

What did we miss by not listening closer?

– Sandeep Chavan

Author of UED, Student of Consequence

About the Author

Sandeep Chavan is an independent researcher, philosopher, and author of over 100 books across science, cosmology, psychology, and metaphysics. A trained engineer turned educator and scientific rebel, he is the founder of the **Universal Energy Dynamics (UED)** framework—a bold and unified reinterpretation of reality based on ripple logic, structured vacuum, and consequence-first thinking.

Known for challenging mainstream physics, rewriting quantum assumptions, and re-centering the observer in all acts of perception, Sandeep's work dismantles dualities and reconstructs science from first principles—without myth, mysticism, or mathematical absolutism.

Through books like *Energy in Action*, *Vishwarup Darshan*, *Manifestation*, and *The War of Thoughts*, he has created a ripple revolution in thought—one that defies classification, but demands attention.

Sandeep lives and teaches in India, where he continues to write, reflect, and ripple with fierce independence, childlike curiosity, and unwavering alignment.

When not rewriting physics, he prefers talking to students under trees, arguing with AI, and losing track of time—because, in his own words, "Time never flowed. I just aligned late."

Don't miss out!

Visit the website below and you can sign up to receive emails whenever Sandeep Chavan publishes a new book. There's no charge and no obligation.

https://books2read.com/r/B-A-EPGPC-TAIQG

BOOKS 2 READ

Connecting independent readers to independent writers.

Did you love *Quantum Courtroom: The Trial of a Century*? Then you should read *Chavan's Post-Materialist Human Model*[1] by Sandeep Chavan!

[2]

What if your thoughts, emotions, actions, and even your fatigue were all just ripples—energetic patterns either aligned or distorted? *This isn't just another self-help book. It's a new model of being.*

You Are Not a Person. You Are a Ripple Becoming Itself.

In **Chavan's Post-Materialist Human Model**, acclaimed thinker and independent researcher Sandeep Chavan

1. https://books2read.com/u/4jjK85

2. https://books2read.com/u/4jjK85

introduces a radical shift in how we understand ourselves—not as static identities or reactive minds, but as dynamic ripple fields. This book reveals the invisible energetic framework behind our decisions, relationships, mental health, and growth—and offers practical, precise tools for tuning it.

Blending insight from neuroscience, emotional regulation, spiritual wisdom, and field-based physics, the book unfolds a fresh paradigm: Alignment is not moral—it's energetic. Clarity is not about intellect—it's ripple literacy. Strength is not force—it's field recovery speed. And personal growth is no longer a collection of habits—it's the art of coherence.

Across fifteen compelling chapters and in-depth appendices, Chavan equips readers to:

Understand ripple noise and why good people often struggleEngineer inner alignment through breath, silence, and micro-pausesRedefine education, therapy, and leadership through ripple intelligenceNavigate relationships without conforming or collapsingCatch the delay between reaction and awareness—and reclaim itReplace diagnosis labels with dynamic energy realignmentLive not for legacy—but as a conscious field shaping it

You'll learn about **Ripple Quotient (RQ)**—a groundbreaking alternative to IQ and EQ, discover practical field-reset rituals, and explore how energy hygiene, sleep, and rhythm directly influence clarity. From trauma healing to everyday decision-making, this book makes the invisible architecture of your life finally visible—and usable.

Whether you're a seeker, therapist, educator, parent, or just someone tired of surface solutions, this book is your invitation to go deeper—to not just improve your life, but *retune* it.

Less motivation. More modulation. Less self-help. More self-alignment. This is *clarity* not as a buzzword, but as a field state. This is *strength* not as toughness, but as ripple resilience.

If you've sensed there's more to your mind than just thoughts—this is your field guide.

Read more at https://www.sandeepjchavan.com/.

Also by Sandeep Chavan

Chavan's UED research

CMGED Research Report

Chavan's Cosmic Continuum (C^3)

Frequency

Ripples of Reality

Reality Realigned

The Great Realignment

I – The Cosmic Leverage

Chavan's Cosmic Theory of Relativity (CTR)

Chavan's Theory of Light

Feeling Like a Physicist

TIME – The New Definition

$E \neq mc^2$: The Collapse of Classical Assumptions

The Chavan Codex

Chavan's Atomic Model

Echoes of War: The Russia-Ukraine Conflict and Its Global Ripples

Shattered Lives

The War Economy
Borders Redrawn
Arms and Allies
Collateral Damage
Justice in the Ashes

Somewhere in the Darkness of the Cosmos
The Blueprint of the Universe
The Cosmic Breath
From the Edge of the Universe

The Eternal Sage Chronicles
The Sage and the Sovereign
The Family Pillar
The People's Flame
The Warrior's Soul
The Bridging Mentor
The Guiding Light

The Exodus
The Last Cycle: 2075

The Fundamental Theory of Life & Existence
Energy in Action: The Soul Demystified

The Active Mind
Energy in Connection
Death: The Disconnection
Subconsciousness: The Chosen Disconnection
Desires: The Engine of Life

The Restoration of Human Knowledge
Restoring Bhagavad Gita

The War of Thoughts
Cracks in the Known
The Human Game
The Divine Dilemma
The Original Intent
The Answers

Ultra Modern Life Philosophy
The Duality of Laughter and Tears
Take It Easy
Compare to Accept

Standalone
It's Not AI, It's AHI - Amplified Human Intelligence
The Decision Paradox

The Four Sapiens
Malicious Script of Indian Polity
The IIT Legacy & Global Impact
The IIT Legacy & Global Impact
You are just a Version of Your Original
Win the Game You Didn't Choose
Always Keep Your Bags Packed
Reveal to Shield - Learn the Game of Facades
Engage Beyond Elections
Culture, Identity & Change: The Evolution of Indian Society
The Triangular Dynamics
Never Trust Without Doubt
If You Can't Change the Country, Leave the Country
Live Fearless with Fear
India: The Investment Magnet
Realignment
Powering a Billion Dreams
Universal Skill
Extension: From Brush to Algorithm
Human Side of Einstein
The Future Knows No Lines
Let's Bring the Generations to the Table
Unified Energy Dynamics
A Beat That Never Dies
Strategic Horizons
Right From Scratch
The Forgotten Innovators
Physics Fixing
The Savior Illusion
The Selfish Side
The Age of the Weakest Sapiens

Neural Wars
The Absorption: There Is No Death
Strange Therapist
Bubbles: Become a Drop
Beyond the Blue
Half-Scholar
The Law of Truth
Failure
Philosophy 2.0: A Structural Manifesto
Chaos: A Gentle Guide to the Unresolvable
Self-Help: New Syllabus
Rippling
Game: How to Play the Share Market
Out of Syllabus
Saffronized: The Code Behind the Crown
The Anger
Moksha: The Liberation
Chavan vs Einstein: The Universe Was Never Curved
Rocket Science: The Ripple Gambling
The Cosmos: The Still Universe
The Taste of Silence
Chavan's Post-Materialist Human Model
10-Second Rule
Chavan vs Hawking
It was Never a Wave Signal
Quantum Courtroom: The Trial of a Century

Watch for more at https://www.sandeepjchavan.com/.

www.ingramcontent.com/pod-product-compliance
Lightning Source LLC
LaVergne TN
LVHW090401210926
848480LV00002B/420